TC 3-21.75
(FM 3-21.75)

The Warrior Ethos and Soldier Combat Skills

AUGUST 2013

DISTRIBUTION RESTRICTION. Approved for public release; distribution is unlimited.

Headquarters Department of the Army

Foreword

Duty, honor, country: Those three hallowed words reverently dictate what you ought to be, what you can be, and what you will be. They are your rallying point to build courage when courage seems to fail, to regain faith when there seems to be little cause for faith, to create hope when hope becomes forlorn.

--- General Douglas MacArthur,
on receiving the Sylvanus Thayer Medal
at the US Military Academy,
12 May, 1962

This publication is available at Army Knowledge Online (https://armypubs.us.army.mil/doctrine/index.html).

To receive publishing updates, please subscribe at http://www.apd.army.mil/AdminPubs/new_subscribe.asp.

*TC 3-21.75 (FM 3-21.75)

Training Circular
No. 3-21.75

Headquarters
Department of the Army
Washington, DC, 13 August 2013

The Warrior Ethos and Soldier Combat Skills

Contents

	Page
Preface	xii

PART ONE. WARRIOR ETHOS

Chapter 1	INTRODUCTION	1-1
	Operational Environment	1-2
	Army Values	1-2
	Law of Land Warfare	1-3
	Warrior Culture	1-3
	Battle Drill	1-3
	Warrior Drills	1-3
Chapter 2	INDIVIDUAL READINESS	2-1
	Predeployment	2-1
	Legal Assistance	2-1
	Personal Weapon	2-3
Chapter 3	COMBAT CARE AND PREVENTIVE MEDICINE	3-1
	Section I. COMBAT CASUALTY CARE	3-1
	Combat Lifesaver	3-1
	Lifesaving Measures (First Aid)	3-2
	Casualty Evacuation	3-28
	Section II. PREVENTIVE MEDICINE	3-38
	Clothing and Sleeping Gear	3-39
	Mental Health and Morale	3-40

Distribution Restriction: Approved for public release; distribution is unlimited.

*This publication supersedes FM 3-21.75, 28 January 2008.

Contents

Chapter 4	ENVIRONMENTAL CONDITIONS	4-1
	Section I. DESERT	4-1
	Types	4-1
	Preparation	4-2
	Section II. JUNGLE	4-5
	Types	4-5
	Preparation	4-7
	Section III. ARCTIC	4-9
	Types	4-9
	Preparation	4-9

PART TWO. SOLDIER COMBAT SKILLS

Chapter 5	COVER, CONCEALMENT, AND CAMOUFLAGE	5-1
	Section I. COVER	5-1
	Natural Cover	5-2
	Man-Made Cover	5-2
	Section II. CONCEALMENT	5-4
	Natural Concealment	5-4
	Actions as Concealment	5-4
	Section III. CAMOUFLAGE	5-5
	Movement	5-5
	Positions	5-5
	Outlines and Shadows	5-5
	Shine	5-5
	Shape	5-5
	Colors	5-6
	Dispersion	5-6
	Preparation	5-6
	Individual Techniques	5-7
Chapter 6	FIGHTING POSITIONS	6-1
	Cover	6-1
	Concealment	6-3
	Camouflage	6-3
	Sectors and Fields of Fire	6-3
	Hasty and Deliberate Fighting Positions	6-4
	Two-Man Fighting Position	6-6
	One-Man Fighting Position	6-17
	Close Combat Missile Fighting Positions	6-20
	Range Cards	6-22
Chapter 7	MOVEMENT	7-1
	Individual Movement Techniques	7-1
	Immediate Actions while Moving	7-5
	Fire and Movement	7-8
	Movement on Vehicles	7-9

Contents

Chapter 8	URBAN AREAS	8-1
	Section I. MOVEMENT TECHNIQUES	8-1
	Avoiding Open Areas	8-1
	Moving Parallel to Buildings	8-1
	Moving Past Windows	8-2
	Crossing a Wall	8-3
	Moving Around Corners	8-4
	Moving Within a Building	8-5
	Section II. OTHER PROCEDURES	8-6
	Entering a Building	8-6
	Clearing a Room	8-10
	Section III. FIGHTING POSITIONS	8-12
	Hasty Fighting Position	8-13
	Prepared Fighting Position	8-15
Chapter 9	'EVERY SOLDIER IS A SENSOR'	9-1
	Definition	9-1
	Resources	9-1
	Forms of Questioning	9-3
	Report Levels	9-3
	SALUTE Format	9-4
	Handling and Reporting of the Enemy	9-5
	Operations Security	9-7
	Observation Techniques	9-8
	Limited Visibility Observation	9-9
	Range Estimation	9-14
Chapter 10	COMBAT MARKSMANSHIP	10-1
	Safety	10-1
	Administrative Procedures	10-1
	Weapons	10-2
	Fire Control	10-14
	Combat Zero	10-16
	Mechanical Zero	10-16
	Battlesight Zero	10-17
	Shot Groups	10-19
	Borelight Zero	10-21
	Misfire Procedures and Immediate Action	10-31
	Reflexive Fire	10-33
Chapter 11	COMMUNICATIONS	11-1
	Section I. MEANS OF COMMUNICATIONS	11-1
	Messengers	11-1
	Wire	11-1
	Visual Signals	11-3
	Sound	11-3
	Radio	11-3

	Section II. RADIOTELEPHONE PROCEDURES	11-4
	Rules	11-4
	Types of Nets	11-4
	Precedence of Reports	11-4
	Message Format	11-5
	Common Messages	11-5
	Prowords	11-6
	Operation on a Net	11-8
	Section III. COMMUNICATIONS SECURITY	11-8
	Classifications	11-8
	Signal Operating Instructions	11-9
	Automated Net Control Device	11-9
	Section IV. EQUIPMENT	11-11
	Radios	11-11
	Wire	11-14
	Telephone Equipment	11-14
Chapter 12	SURVIVAL, EVASION, RESISTANCE, AND ESCAPE	12-1
	Survival	12-1
	Evasion	12-4
	Resistance	12-6
	Escape	12-8
Chapter 13	CHEMICAL, BIOLOGICAL, RADIOLOGICAL, OR NUCLEAR WEAPONS	13-1
	Section I. CHEMICAL WEAPONS	13-1
	Types	13-1
	Detection	13-4
	Protective Actions	13-8
	Protective Equipment	13-8
	Mission-Oriented Protective Posture	13-10
	Decontamination	13-10
	Section II. BIOLOGICAL WEAPONS	13-14
	Types	13-14
	Detection	13-14
	Decontamination	13-14
	Protection	13-14
	Section III. RADIOLOGICAL WEAPONS	13-15
	Types	13-15
	Detection	13-16
	Decontamination	13-16
	Protection	13-16
	Section IV. NUCLEAR WEAPONS	13-17
	Characteristics	13-17
	Detection	13-18
	Decontamination	13-19
	Protection	13-20

Chapter 14	MINES, DEMOLITIONS, AND BREACHING PROCEDURES	14-1
	Section I. MINES	14-1
	Antipersonnel Mines	14-2
	M21 Antitank Mine	14-7
	Section II. DEMOLITION FIRING SYSTEMS	14-8
	Booster Assemblies	14-8
	Misfires	14-13
	Section III. OBSTACLES	14-14
	Breach and Cross a Minefield	14-14
	Breach and Cross a Wire Obstacle	14-17
Chapter 15	UXO AND IEDS	15-1
	Section I. UNEXPLODED ORDNANCE	15-1
	Dropped Ordnance	15-1
	Projected Ordnance	15-6
	Thrown Ordnance (Hand Grenades)	15-7
	Section II. IMPROVISED EXPLOSIVE DEVICES	15-9
	Types	15-9
	Identification	15-10
	Components	15-10
	Examples	15-12
	Actions on Finding UXO	15-15
	Actions on Finding IEDs	15-16
Appendix A	CHECKLISTS AND MEMORY AIDS	A-1
Glossary		Glossary-1
References		References-1
Index		Index-1

Figures

Figure 1-1. Army Values	1-2
Figure 1-2. Warrior drills	1-4
Figure 2-1. Example personal predeployment checklist	2-2
Figure 3-1. Assessment	3-3
Figure 3-2. Airway blocked by tongue	3-4
Figure 3-3. Airway opened by extending neck	3-4
Figure 3-4. Jaw-thrust technique	3-5
Figure 3-5. Head-tilt/chin-lift technique	3-5
Figure 3-6. Check for breathing	3-6
Figure 3-7. Rescue breathing	3-7
Figure 3-8. Placement of fingers to detect pulse	3-8

Contents

Figure 3-9. Abdominal thrust on unresponsive casualty. 3-10
Figure 3-10. Hand placement for chest thrust. 3-11
Figure 3-11. Breastbone depressed 1 1/2 to 2 inches. 3-11
Figure 3-12. Opening of casualty's mouth, tongue-jaw lift. 3-12
Figure 3-13. Opening of casualty's mouth, crossed-finger method. 3-12
Figure 3-14. Use of finger to dislodge a foreign body. 3-13
Figure 3-15. Emergency bandage. 3-14
Figure 3-16. Application of pad to wound. 3-15
Figure 3-17. Insertion of bandage into pressure bar. 3-15
Figure 3-18. Tightening of bandage. 3-15
Figure 3-19. Pressure of bar into bandage. 3-15
Figure 3-20. Wrapping of bandage over pressure bar. 3-16
Figure 3-21. Securing of bandage. 3-16
Figure 3-22. Grasping of dressing tails with both hands. 3-16
Figure 3-23. Pulling open of dressing. 3-17
Figure 3-24. Placement of dressing directly on wound. 3-17
Figure 3-25. Wrapping of dressing tail around injured part. 3-17
Figure 3-26. Tails tied into nonslip knot. 3-17
Figure 3-27. Application of direct manual pressure. 3-17
Figure 3-28. Elevation of injured limb. 3-18
Figure 3-29. Wad of padding on top of field dressing. 3-19
Figure 3-30. Improvised dressing over wad of padding. 3-19
Figure 3-31. Ends of improvised dressing wrapped tightly around limb. 3-19
Figure 3-32. Ends of improvised dressing tied together in nonslip knot. 3-19
Figure 3-33. Digital pressure (fingers, thumbs, or hands). 3-20
Figure 3-34. Band pulled tight. 3-22
Figure 3-35. Improved first aid kit. 3-23
Figure 3-36. Tourniquet above knee. 3-24
Figure 3-37. Rigid object on top of half knot. 3-24
Figure 3-38. Tourniquet knotted over rigid object and twisted. 3-25
Figure 3-39. Free ends tied on side of limb. 3-25
Figure 3-40. Fireman's carry. 3-30
Figure 3-41. Alternate fireman's carry. 3-32
Figure 3-42. Supporting carry. 3-33
Figure 3-43. Neck drag. 3-34
Figure 3-44. Cradle drop drag. 3-35
Figure 3-45. Two-man support carry. 3-36
Figure 3-46. Two-man fore-and-aft carry. 3-37
Figure 3-47. Two-hand seat carry. 3-37
Figure 3-48. Rules for avoiding illness in the field. 3-39
Figure 3-49. Care of the feet. 3-40
Figure 5-1. Natural cover. 5-1

Figure 5-2. Cover along a wall ... 5-2
Figure 5-3. Man-made cover ... 5-2
Figure 5-4. Body armor and helmet ... 5-3
Figure 5-5. Protective cover against chemical/biological warfare agents ... 5-3
Figure 5-6. Concealment ... 5-4
Figure 5-7. Soldier in arctic camouflage ... 5-6
Figure 5-8. Camouflaged Soldiers ... 5-7
Figure 5-9. Camouflaged helmet ... 5-8
Figure 5-10. Advanced camouflage face paint ... 5-9
Figure 6-1. Man-made cover ... 6-1
Figure 6-2. Cover ... 6-2
Figure 6-3. Prone position (hasty) ... 6-6
Figure 6-4. Establishment of sectors and building method ... 6-8
Figure 6-5. Two-man fighting position (Stage 1) ... 6-9
Figure 6-6. Placement of OHC supports and construction of retaining walls ... 6-10
Figure 6-7. Two-man fighting position (Stage 2) ... 6-10
Figure 6-8. Digging of position (side view) ... 6-11
Figure 6-9. Placement of stringers for OHC ... 6-11
Figure 6-10. Two-man fighting position (Stage 3) ... 6-12
Figure 6-11. Revetment construction ... 6-12
Figure 6-12. Grenade sumps ... 6-13
Figure 6-13. Storage compartments ... 6-13
Figure 6-14. Installation of overhead cover ... 6-14
Figure 6-15. Two-man fighting position with built-up OHC (Stage 4) ... 6-15
Figure 6-16. Two-man fighting position with built-down OHC (top view) ... 6-16
Figure 6-17. Two-man fighting position with built-down OHC (side view) ... 6-17
Figure 6-18. Position with firing platforms ... 6-18
Figure 6-19. Grenade sump locations ... 6-19
Figure 6-20. Machine gun fighting position with OHC ... 6-20
Figure 6-21. Standard Javelin fighting position ... 6-21
Figure 6-22. Primary sector with an FPL ... 6-24
Figure 6-23. Complete sketch with PDF ... 6-25
Figure 6-24. Data section ... 6-26
Figure 6-25. Example completed data section ... 6-27
Figure 6-26. Example completed range card ... 6-28
Figure 6-27. Reference points and target reference points ... 6-30
Figure 6-28. Maximum engagement lines ... 6-31
Figure 6-29. Weapon reference point ... 6-32
Figure 7-1. Low and high crawl ... 7-2
Figure 7-2. Rush ... 7-3
Figure 7-3. Fire team wedge ... 7-5
Figure 7-4. Following of team leader from impact area ... 7-6

Figure 7-5. Reaction to ground flares. ... 7-7
Figure 7-6. Reaction to aerial flares. ... 7-8
Figure 7-7. Mounting and riding arrangements. ... 7-10
Figure 8-1. Soldier moving past windows. ... 8-2
Figure 8-2. Soldier passing basement windows. ... 8-3
Figure 8-3. Soldier crossing a wall. ... 8-3
Figure 8-4. Correct technique for looking around a corner. ... 8-4
Figure 8-5. *Pie-ing* a corner. ... 8-4
Figure 8-6. Movement within a building. ... 8-5
Figure 8-7. Lower-level entry technique with support bar. ... 8-7
Figure 8-8. Lower-level entry technique without support bar. ... 8-7
Figure 8-9. Lower-level entry two-man pull technique. ... 8-8
Figure 8-10. Lower-level entry one-man lift technique. ... 8-8
Figure 8-11. M433 HEDP grenade. ... 8-10
Figure 8-12. Some considerations for selecting and occupying individual fighting positions ... 8-12
Figure 8-13. Soldier firing left or right handed. ... 8-13
Figure 8-14. Soldier firing around a corner. ... 8-14
Figure 8-15. Soldier firing from peak of a roof. ... 8-15
Figure 8-16. Emplacement of machine gun in a doorway. ... 8-17
Figure 9-1. Potential indicators. ... 9-2
Figure 9-2. Example captured document tag. ... 9-6
Figure 9-3. Rapid/slow-scan pattern. ... 9-8
Figure 9-4. Detailed search. ... 9-9
Figure 9-5. Typical scanning pattern. ... 9-10
Figure 9-6. Off-center viewing. ... 9-11
Figure 9-7. AN/PVS-7 and AN/PVS-14. ... 9-12
Figure 9-8. AN/PAS-13, V1, V2, and V3. ... 9-13
Figure 9-9. AN/PAQ-4-series and the AN/PEQ-2A. ... 9-13
Figure 9-10. Mil-relation formula. ... 9-16
Figure 10-1. M9 pistol. ... 10-3
Figure 10-2. M16A2 rifle. ... 10-4
Figure 10-3. M4 carbine. ... 10-5
Figure 10-4. M203 grenade launcher. ... 10-6
Figure 10-5. M249 squad automatic weapon (SAW). ... 10-7
Figure 10-6. M240B machine gun. ... 10-8
Figure 10-7. M2 .50 caliber machine gun with M3 tripod mount. ... 10-9
Figure 10-8. MK 19 grenade machine gun, Mod 3. ... 10-10
Figure 10-9. Improved M72 LAW. ... 10-11
Figure 10-10. M136 AT4. ... 10-12
Figure 10-11. M141 BDM. ... 10-13
Figure 10-12. Javelin. ... 10-14
Figure 10-13. M16A2/A3 rifle mechanical zero. ... 10-17

Figure 10-14. M16A4 and M4 carbine rifle mechanical zero. ... 10-17
Figure 10-15. M16A2/A3 rifle battlesight zero. ... 10-18
Figure 10-16. M16A4 rifle battlesight zero. ... 10-18
Figure 10-17. M4 rifle battlesight zero. ... 10-19
Figure 10-18. Final shot group results. ... 10-20
Figure 10-19. Example zeroing mark. ... 10-22
Figure 10-20. Borelight in the start point position. ... 10-23
Figure 10-21. Borelight in the half-turn position. ... 10-23
Figure 10-22. Examples of start point, half-turn, and reference point. ... 10-24
Figure 10-23. Blank 10-meter boresight target and offset symbols. ... 10-25
Figure 10-24. Backup iron sight. ... 10-25
Figure 10-25. M68 close combat optic. ... 10-26
Figure 10-26. Example TWS zeroing adjustments. ... 10-28
Figure 10-27. Example shot group adjustment with strike zone. ... 10-29
Figure 10-28. Ready positions. ... 10-34
Figure 11-1. Common prowords. ... 11-7
Figure 11-2. Automated net control device. ... 11-9
Figure 11-3. Automated net control device keypad. ... 11-9
Figure 11-4. Call signs. ... 11-10
Figure 11-5. Time periods. ... 11-11
Figure 11-6. AN/PRC-148 multiband intrateam radio (MBITR). ... 11-12
Figure 11-7. IC-F43 portable UHF transceiver. ... 11-12
Figure 11-8. AN/PRC-119A-D SIP. ... 11-13
Figure 11-8. AN/PRC-119E advanced system improvement program (ASIP). ... 11-13
Figure 12-1. SURVIVAL. ... 12-2
Figure 12-2. Tool for remembering shelter locations. ... 12-5
Figure 12-3. Code of Conduct. ... 12-7
Figure 13-1. CANA. ... 13-2
Figure 13-2. NAAK, Mark I. ... 13-2
Figure 13-3. M22 ACADA. ... 13-5
Figure 13-4. M22 ICAM. ... 13-5
Figure 13-5. M8 chemical agent detector paper. ... 13-6
Figure 13-6. M9 chemical agent detector paper. ... 13-6
Figure 13-7. M256 chemical agent detector kit. ... 13-7
Figure 13-8. Protective mask M40A1/A2 and M42A2 CVC. ... 13-9
Figure 13-9. M291 skin decontaminating kit. ... 13-13
Figure 13-10. M295 equipment decontamination kit. ... 13-13
Figure 13-11. M100 SORBENT Decontamination System. ... 13-13
Figure 13-12. Radiac Set AN/VDR 2. ... 13-18
Figure 13-13. Radiac set AN/UDR 13. ... 13-19
Figure 14-1. M18A1 antipersonnel mine. ... 14-3

Contents

Figure 14-2. M7 bandoleer. ... 14-3
Figure 14-3. M18A1 antipersonnel mine data. ... 14-4
Figure 14-4. M-131 Modular Pack Mine System (MOPMS). ... 14-6
Figure 14-5. M-131 MOPMS deployed. ... 14-6
Figure 14-6. MOPMS emplacement and safety zone. ... 14-7
Figure 14-7. M21 antitank mine and components. ... 14-8
Figure 14-8. MDI components. ... 14-9
Figure 14-9. Priming of C4 demolition blocks with MDI. ... 14-10
Figure 14-10. Priming of C4 demolition blocks with detonating cord. ... 14-10
Figure 14-11. Priming of C4 with L-shaped charge. ... 14-11
Figure 14-12. Preparation of M81 fuse igniter. ... 14-11
Figure 14-13. M81 fuse igniter with the M14 time fuse delay. ... 14-12
Figure 14-14. M81 fuse igniter with the M9 holder. ... 14-12
Figure 14-15. M81 fuse igniter with the M14 time fuse delay. ... 14-13
Figure 14-16. Mine probe. ... 14-15
Figure 14-17. Lanes. ... 14-15
Figure 14-18. Knot toward mine. ... 14-16
Figure 14-19. Marked mines. ... 14-16
Figure 14-20. Bangalore torpedo. ... 14-17
Figure 14-21. Priming of bangalore torpedo with MDI. ... 14-18
Figure 14-22. MK7 Antipersonnel Obstacle-Breaching System (APOBS). ... 14-18
Figure 15-1. Antipersonnel, ball-type submunitions. ... 15-3
Figure 15-2. Area-denial submunitions (conventional). ... 15-3
Figure 15-3. Antipersonnel/AMAT submunitions (conventional). ... 15-4
Figure 15-4. AMAT/antitank submunitions (conventional). ... 15-5
Figure 15-5. Fragmentation grenades. ... 15-8
Figure 15-6. Antitank grenades. ... 15-8
Figure 15-7. Smoke grenades. ... 15-8
Figure 15-8. U.S. illumination grenade. ... 15-8
Figure 15-9. Vehicle IED capacities and danger zones. ... 15-10
Figure 15-10. Main charge (explosives). ... 15-11
Figure 15-11. Casing (material around the explosives). ... 15-11
Figure 15-12. Initiators (command detonated, victim activated, with timer). ... 15-12
Figure 15-13. IED components. ... 15-13
Figure 15-14. IED transmitters and receivers. ... 15-13
Figure 15-15. Common objects as initiators. ... 15-14
Figure 15-16. Unexploded rounds as initiators. ... 15-14
Figure 15-17. Emplaced IED with initiator. ... 15-15
Figure 15-18. Electric blasting caps. ... 15-15
Figure 15-19. Nine-Line UXO Incident Report. ... 15-16
Figure 15-20. IED Spot Report. ... 15-17

Tables

Table 3-1. First aid. ... 3-2
Table 5-1. Application of camouflage face paint to skin. ... 5-9
Table 6-1. Characteristics of individual fighting positions. ... 6-5
Table 6-2. Construction of two-man fighting position. ... 6-7
Table 6-3. Specifications for built-down overhead cover. ... 6-16
Table 9-1. SALUTE format line by line. ... 9-5
Table 9-2. Appearance of a body using appearance-of-objects method. ... 9-16
Table 11-1. Comparison of communication methods. ... 11-2
Table 13-1. MOPP levels. ... 13-10
Table 13-2. Decontamination levels and techniques. ... 13-12

Preface

This training circular provides all Soldiers the doctrinal basis for the Warrior Ethos, Warrior Tasks, and other combat-critical tasks. It also updates weapon, equipment, and munitions information. This FM is not intended to serve as a stand-alone publication. It should be used with other Army publications that contain more in-depth information.

The target audience for this publication includes individual Soldiers and noncommissioned officers throughout the Army.

This book applies to the Active Army, the Army National Guard (ARNG)/National Guard of the United States (ARNGUS), and the U.S. Army Reserve (USAR) unless otherwise stated.

The proponent for this publication is the U.S. Army Training and Doctrine Command (TRADOC). The preparing agency is the U.S. Army Maneuver Center of Excellence (MCoE). Send comments and recommendations by any means, U.S. mail, e-mail, fax, or telephone, using the format of DA Form 2028, *Recommended Changes to Publications and Blank Forms*. Point of contact information is as follows.

E-mail:	curtis.d.archuleta.civ@mail.mil
Phone:	COM 706-545-7114 or DSN 835-7114
Fax:	COM 706-545-8511 or DSN 835-8511
U.S. Mail:	Commanding General, MCoE
	Directorate of Training and Doctrine (DOTD)
	Doctrine and Collective Training Division
	ATTN: ATZB-TDD
	Fort Benning, GA 31905-5410

Uniforms shown in this training circular were drawn without camouflage for clarity of the illustration.

Unless this publication states otherwise, masculine nouns and pronouns may refer to either men or women.

Introduction

Modern combat is chaotic, intense, and shockingly destructive. In your first battle, you will experience the confusing and often terrifying sights, sounds, smells, and dangers of the battlefield—but you must learn to survive and win despite them.

1. You could face a fierce and relentless enemy.
2. You could be surrounded by destruction and death.
3. Your leaders and fellow soldiers may shout urgent commands and warnings.
4. Rounds might impact near you.
5. The air could be filled with the smell of explosives and propellant.
6. You might hear the screams of a wounded comrade.

However, even in all this confusion and fear, remember that you are not alone. You are part of a well-trained team, backed by the most powerful combined arms force, and the most modern technology in the world. You must keep faith with your fellow Soldiers, remember your training, and do your duty to the best of your ability. If you do, and you uphold your Warrior Ethos, you can win and return home with honor.

This is the Soldier's TC. It tells the Soldier how to perform the combat skills needed to survive on the battlefield. All Soldiers, across all branches and components, must learn these basic skills. Noncommissioned officers (NCOs) must ensure that their Soldiers receive training on--and know—these vital combat skills.

This page intentionally left blank.

PART ONE

Warrior Ethos

What is Warrior Ethos? At first glance, it is just four simple lines embedded in the Soldier's Creed. Yet, it is the spirit represented by these four lines that--

- Compels Soldiers to fight through all adversity, under any circumstances, in order to achieve victory.
- Represents the US Soldier's loyal, tireless, and selfless commitment to his nation, his mission, his unit, and his fellow Soldiers.
- Captures the essence of combat, Army Values, and Warrior Culture.

Sustained and developed through discipline, commitment, and pride, these four lines motivate every Soldier to persevere and, ultimately, to refuse defeat. These lines go beyond mere survival. They speak to forging victory from chaos; to overcoming fear, hunger, deprivation, and fatigue; and to accomplishing the mission:

THE SOLDIER'S CREED

I am an American Soldier.

I am a Warrior and a member of a team.

I serve the people of the United States and live the Army Values.

> I will always place the mission first.
> I will never accept defeat.
> I will never quit.
> I will never leave a fallen comrade.

I am disciplined, physically and mentally tough, trained and proficient in my Warrior tasks and drills.

I always maintain my arms, my equipment, and myself.

I am an expert and I am a professional.

I stand ready to deploy, engage, and destroy the enemies of the United States of America in close combat.

I am a guardian of freedom and the American way of life.

I am an American Soldier.

Chapter 1

Introduction

Military service is more than a "job." It is a profession with the enduring purpose to win wars and destroy our nation's enemies. The Warrior Ethos demands a dedication

Chapter 1

to duty that may involve putting your life on the line, even when survival is in question, for a cause greater than yourself. As a Soldier, you must motivate yourself to rise above the worst battle conditions—no matter what it takes, or how long it takes. That is the heart of the Warrior Ethos, which is the foundation for your commitment to victory in times of peace and war. While always exemplifying the four parts of Warrior Ethos, you must have absolute faith in yourself and your team, as they are trained and equipped to destroy the enemy in close combat. Warrior drills are a set of nine battle drills, consisting of individual tasks that develop and manifest the Warrior Ethos in Soldiers.

OPERATIONAL ENVIRONMENT

1-1. This complex operational environment offers no relief or rest from contact with the enemy across the spectrum of conflict. No matter what combat conditions you find yourself in, you must turn your personal Warrior Ethos into your commitment to win. In the combat environment of today, unlike conflicts of the past, there is little distinction between the forward and rear areas. Battlefields of the Global War on Terrorism, and battles to be fought in the US Army's future, are and will be asymmetrical, violent, unpredictable, and multidimensional. Today's conflicts are fought throughout the whole spectrum of the battlespace by all Soldiers, regardless of military occupational specialty (MOS). Every Soldier must think as a Warrior first; a professional Soldier, trained, ready, and able to enter combat; ready to fight--and win--against *any* enemy, *any* time, *any* place.

ARMY VALUES

1-2. US Army Values reminds us and displays to the rest of the world—the civilian governments we serve, the nation we protect, other nations, and even our enemies—who we are and what we stand for (Figure 1-1). The trust you have for your fellow Soldiers, and the trust the American people have in you, depends on how well you live up to the Army Values. After all, these values are the fundamental building blocks that enable you to understand right from wrong in any situation. Army Values are consistent and support one another; you cannot follow one value and ignore the others. Figure 1-1 shows the Army Values, which form the acrostic LDRSHIP.

Loyalty	Bear true faith and allegiance to the Constitution, the Army, your unit, and other Soldiers.
Duty	Fulfill your obligations.
Respect	Treat people with dignity as they should be treated.
Selfless Service	Put the welfare of the nation, the Army, and your subordinates before your own.
Honor	Live up to all the Army Values.
Integrity	Do what's right, legally and morally.
Personal Courage (Physical or Moral)	Face fear, danger, or adversity.

Figure 1-1. Army Values.

1-3. Performance in combat, the greatest challenge, requires a basis, such as Army Values, for motivation and will. In these values are rooted the basis for the character and self-discipline that generates the will to succeed and the motivation to persevere. From this motivation derived through tough realistic training and the skills acquired, which will make you successful, a Soldier who "walks the walk."

1-4. Army Values, including policies and procedures, form the foundation on which the Army's institutional culture stands. However, written values are useless unless practiced. You must act correctly with character, complete understanding, and sound motivation. Your trusted leaders will aid you in adopting such values by making sure their core experiences validate them. By this method, strategic leadership embues Army Values into all Soldiers.

LAW OF LAND WARFARE

1-5. The conduct of armed hostilities on land is regulated by FM 27-10 and the Law of Land Warfare. Their purpose is to diminish the evils of war by protecting combatants *and* noncombatants from unnecessary suffering, and by safeguarding certain fundamental human rights of those who fall into the hands of the enemy, particularly enemy prisoners of war (EPWs), detainees, wounded and sick, and civilians. Every Soldier adheres to these laws, and ensures that his subordinates adhere to them as well, during the conduct of their duties. Soldiers must also seek clarification from their superiors of any unclear or apparently illegal order. Soldiers need to understand that the law of land warfare not only applies to states, but also to individuals, particularly all members of the armed forces.

WARRIOR CULTURE

1-6. The Warrior Culture, a shared set of important beliefs, values, and assumptions, is crucial and perishable. Therefore, the Army must continually affirm, develop, and sustain it, as it maintains the nation's existence. Its martial ethic connects American warriors of today with those whose previous sacrifices allowed our nation to persevere. You, the individual Soldier, are the foundation for the Warrior Culture. As in larger institutions, the Armed Forces' use culture, in this case Warrior Culture, to let people know they are part of something bigger than just themselves: they have responsibilities not only to the people around them, but also to those who have gone before and to those who will come after them. The Warrior Culture is a part of who you are, and a custom you can take pride in. Personal courage, loyalty to comrades, and dedication to duty are attributes integral to putting your life on the line.

BATTLE DRILL

1-7. A battle drill--

- Is a collective action, executed by a platoon or smaller element, without the application of a deliberate decision-making process. The action is vital to success in combat or critical to preserve life. The drill is initiated on a cue, such as an enemy action or your leader's order, and is a trained response to the that stimulus. It requires minimum leader orders to accomplish, and is standard throughout the Army. A drill has the following advantages:
  -- It is based on unit missions and the specific tasks, standards, and performance measures required to support mission proficiency.
  -- It builds from simple to complex, but focuses on the basics.
  -- It links how-to-train and how-to-fight at small-unit level.
  -- It provides an agenda for continuous coaching and analyzing.
  -- It develops leaders, and builds teamwork and cohesion under stress.
  -- It enhances the chance for individual and unit survival on the battlefield.

WARRIOR DRILLS

1-8. The Warrior drills--

- Are a set of core battle drills for small units from active and reserve component organizations across the Army, regardless of branch.
- Describe a training method for small units. This method requires training individual, leader, and collective tasks before the conduct of critical wartime missions.
- Provide a foundation for the development of specific objectives for combat. The expanded list of Warrior Drills helps place the individual Soldiers' tasks (as well as the team) in sufficient context to identify meaningful consequences of individual behavior.
- Have individual tasks that develop and manifest the Warrior Ethos. A *barrier*, for example, is an element that impedes a response or behavior. *Barrier control* can focus on points that are most sensitive to the behavior of individuals such as choices, actions, and interactions; and on those with the most serious consequences such as effects on other individuals and success of the mission.
- Create opportunities to develop the Warrior Ethos. The nine drills follow:

---

1. React to Contact (visual, improvised explosive device [IED], direct fire).

2. React to Ambush (Near).

3. React to Ambush (Far).

4. React to Indirect Fire.

5. React to a Chemical Attack.

6. Break Contact.

7. Dismount a Vehicle.

8. Evacuate Wounded Personnel from Vehicle.

9. Establish Security at the Halt.

Figure 1-2. Warrior drills.

Chapter 2
Individual Readiness

The US Army is based on our nation's greatest resource-you, the individual fighting Soldier. Success in the defense of our nation depends on your individual readiness, initiative, and capabilities. You are cohesive, integral parts of the whole. Your mission is to deter aggression through combat readiness and, when deterrence fails, to win the nation's wars. This mission must not be compromised. You must be ready.

Deployment is challenging and stressful--both on you and on your family. You will be away from the comforts of home. This is not easy. Preparedness can reduce the stress and increase your focus and confidence once you are deployed.

PREDEPLOYMENT

2-1. What could or would happen if you were a long way from your family for an indefinite period of time, and unable to communicate with them? The losing organization will complete a DA Form 7425, *Readiness and Deployment Checklist*, on you, but it is not designed for your use. Figure 2-1 shows an example checklist that you might create for your own use.

LEGAL ASSISTANCE

2-2. Your installation legal assistance center can provide a great number of services. Army legal assistance centers provide answers and advice to even the most complex problems. Such legal assistance usually does not include in-court representation. Some of the issues that your installation's legal assistance center may be able to help with follow:

- Marriage and divorce issues.
- Child custody and visitation issues.
- Adoptions or other family matters (as expertise is available).
- Wills.
- Powers of attorney.
- Advice for designating SGLI beneficiaries.
- Landlord-tenant issues.
- Consumer affairs such as mortgages, warranties.
- Bankruptcies.
- Garnishments and indebtedness.
- Notarizations.
- Name changes, as expertise is available.
- Bars to reenlistment (as available).
- Hardship discharges.
- Taxes.

Chapter 2

Defense Enrollment Eligiblty Reporting System (DEERS)	Verify your DEERS information and ensure your family members can get needed medical care in your absence.
Dental Records	Update if needed.
Documents, Locations of	Ensure that your spouse or other family member(s) know where to find all of the above documents.
DD Form 93 Emergency Data Record	Check to ensure this is current and correct.
Eyeglasses and Protective Mask Inserts	Ensure you have two pairs of eyeglasses and protective mask inserts, all with your current prescription, if required.
Family Assistance Army Community Service (ACS)	Tell your family where to get various kinds of support and help while you are gone.
Legal Aid, Military	Tell your spouse where to get military legal aid in your absence.
Readiness Group (FRG)	Tell your family where to get various kinds of support and help while you are gone.
Finances	Ensure your spouse has access to all of your records and accounts and update them as needed.
Identification	Ensure you have two sets of these, if required.
Legal	See Family Assistance.
Life Insurance	Designate your beneficiary on SGLV Forms 8286 and 8286A, Soldier's Group Life Insurance (SGLI) Election and Certificate.
Directive, Advance	Specify any decisions you wish others to make on your behalf should you be unable to do so for yourself.
Medical (see also Power of Attorney) DD Form 2766 (Shot Record)	Keep your vaccinations and immunizations current.
Directive, Advance	Prepare if you want to specify how decisions are made on your behalf should you be unable to do so for yourself.
Living Will	Prepare if desired.
Records	Update if needed.
Warnings Tags	Ensure you have two sets of these, if required.
Power of Attorney General	Prepare to allow someone to perform all duties for you in your absence.
Medical, Durable	Prepare one of these to designate who makes decisions for you or your dependents, including your minor children, should a medical emergency occur while you are deployed or otherwise unable to make the decision yourself.
Special	Prepare to allow someone to perform a particular kind of duty for you in your absence.
Property	Prepare or update accounts, documents, and records as needed.
Service Record	Check to ensure this is current and correct.
Training	Update your weapon qualification(s), if needed.
Will(s)	Prepare new or update existing, for you and your spouse, if needed.

Figure 2-1. Example personal predeployment checklist.

PERSONAL WEAPON

2-3. Your personal weapon is vital to you in combat. Take care of it, and it will do the same for you. Seems obvious, right? Apparently not. Multiple reports from the opening days of Operation Iraqi Freedom revealed that faulty weapons training and maintenance were the main causes of US casualties and captures: "These malfunctions may have resulted from inadequate individual maintenance and the environment." Soldiers had trouble firing their personal and crew-served weapons, and the main reason cited was poor preventive maintenance. Few things will end a firefight faster and more badly than a weapon that will not shoot! The complex M16A2 rifle needs cleaning and proper lubrication at least once a day in order to properly function. Follow these procedures and those in the technical manual (TM):

CLEANING

2-4. Use only the cleaning supplies listed in the *Expendable and Durable Items List* in the back of the TM.

Abrasives and Harsh Chemicals

2-5. Avoid using abrasive materials such as steel wool or commercial scrubbing pads, and harsh chemicals not intended for use on your weapon. This can ruin the finish of the weapon. It can also remove rifling and damage internal parts, either of which can make your weapon inaccurate and ineffective during the mission.

Water

2-6. Never clean your weapon under running water, which can force moisture into tight places, resulting in corrosion.

Frequency

2-7. In the field, clean your weapon often, at least daily. Even just taking every chance to wipe the weapon's exterior with a clean cloth will help ensure operability.

Disassembly

2-8. Do any cleaning that involves disassembly at your level in an enclosed area. Blowing sand and other debris can not only affect your weapon, it can also cause you to lose the parts of the weapon. For parts that must be disassembled beyond your level, such as the trigger assembly, just blow out the dirt or debris.

Magazines and Ammunition

2-9. Clean your magazines, but avoid using any lubrication in them or on ammunition. Unload and wipe off your ammunition daily, then disassemble and run a rag through the magazine to prevent jamming.

LUBRICATING

2-10. Lubrication reduces friction between metal parts.

Lubricant

2-11. You may only use authorized, standard military lubricant for small arms such as cleaner lubricant preservative (CLP). Also, lubricate only internal parts.

Chapter 2

Moving Parts

2-12. Pay special attention to moving parts like the bolt carrier. Wipe the outside of the weapon dry.

Covers and Caps

2-13. Use rifle covers and muzzle caps to keep blowing debris and dust out of the muzzle and ejection port area. Cover mounted machine guns when possible. Keep your rifle's ejection port cover closed and a magazine inserted.

Humid Environments

2-14. Keep in mind that, in more humid environments such as jungles and swamps, you will need to use more lubrication, more often, on all metal parts. Temperature and other extreme weather conditions also factor in.

Desert Environments

2-15. Corrosion poses little threat in the desert. Avoid using too much lubrication, because it attracts sand.

Note: Maintain all issued equipment and clothing based on the specific care and maintenance instructions provided.

Chapter 3

Combat Casualty Care and Preventive Medicine

Combat casualty care is the treatment administered to a wounded Soldier after he has been moved out of an engagement area or the enemy has been suppressed. This level of care can help save life and limb until medical personnel arrive. Soldiers might have to depend upon their own first-aid knowledge and skills to save themselves (self-aid) or another Soldier (buddy aid or combat lifesaver skills). This knowledge and training can possibly save a life, prevent permanent disability, or reduce long periods of hospitalization. The only requirement is to know what to do--and what not to do--in certain instances.

Personal hygiene and preventive medicine are simple, common-sense measures that each Soldier can perform to protect his health and that of others. Taking these measures can greatly reduce time lost due to disease and nonbattle injury.

Section I. COMBAT CASUALTY CARE

The Army warfighter doctrine, developed for a widely dispersed and rapidly moving battlefield, recognizes that battlefield constraints limit the number of trained medical personnel available to provide immediate, far-forward care. This section defines combat lifesaver, provides life-saving measures (first aid) techniques, and discusses casualty evacuation.

COMBAT LIFESAVER

3-1.   The role of the combat lifesaver was developed to increase far-forward care to combat Soldiers. At least one member--though ideally *every* member of each squad, team, and crew--should be a trained combat lifesaver. The leader is seldom a combat lifesaver, since he will have less time to perform those duties than would another member of his unit.

3-2.   So what exactly is a combat lifesaver? He is a nonmedical combat Soldier. His *secondary* mission is to help the combat medic provide basic emergency care to injured members of his squad, team, or crew, and to aid in evacuating them, mission permitting. He complements, rather than replaces, the combat medic. He receives training in enhanced first aid and selected medical procedures such as initiating intravenous infusions. Combat lifesaver training bridges the first aid training (self-aid or buddy aid, or SABA) given to all Soldiers in basic training, and the more advanced medical training given to Medical Specialists (MOS 91W), also known as combat medics.

3-3.   The Academy of Health Sciences developed the Combat Lifesaver Course as part of its continuing effort to provide health service support to the Army. The current edition of the Combat Lifesaver Course lasts three days. The first day tests the buddy-aid tasks, and the other two days teach and test specific medical tasks.

LIFESAVING MEASURES (FIRST AID)

3-4. When a Soldier is wounded, he must receive first aid immediately. Most injured or ill Soldiers can return to their units to fight or support. This is mainly, because they receive appropriate and timely first aid, followed by the best possible medical care. To help ensure this happens, every Soldier should have combat lifesaver training on basic life-saving procedures (Table 3-1).

Table 3-1. First aid.

1	Check for BREATHING	Lack of oxygen, due either to a compromised airway or inadequate breathing, can cause brain damage or death in just a few minutes.
2	Check for BLEEDING	Life can continue only with sufficient blood to carry oxygen to tissues.
3	Check for SHOCK	Unless shock is prevented, first aid performed, and medical treatment provided, death may result, even with an otherwise nonfatal injury.

CHECK FOR BREATHING

3-5. Check first to see if the casualty's heart is beating, then to see if he is breathing. This paragraph discusses what to do in each possible situation.

React to Stoppage of Heartbeat

3-6. If a casualty's heart stops beating, you must immediately seek medical help. *Seconds count!* Stoppage of the heart is soon followed by cessation of respiration, unless that has already happened. Remain calm, but think first, and act quickly. When a casualty's heart stops, he has no pulse. He is unconscious and limp, and his pupils are open wide. When evaluating a casualty, or when performing the preliminary steps of rescue breathing, feel for a pulse. If you *do not* detect a pulse, seek medical help.

Open Airway and Restore Breathing

3-7. All humans need oxygen to live. Oxygen breathed into the lungs gets into the bloodstream. The heart pumps the blood, which carries the oxygen throughout the body to the cells, which require a constant supply of oxygen. Without a constant supply of oxygen to the cells in the brain, we can suffer permanent brain damage, paralysis, or death.

Assess and Position Casualty

3-8. To assess the casualty, do the following:

1. *Check* for responsiveness (A, Figure 3-1). Establish whether the casualty is conscious by gently shaking him and asking, "Are you OK?"
2. *Call* for help, if appropriate (B, Figure 3-1).
3. *Position* the unconscious casualty so that he is lying on his back and on a firm surface (C, Figure 3-1).

WARNING

If the casualty is lying on his chest (prone), cautiously roll him as a unit, so that his body does not twist. Twisting him could complicate a back, neck, or spinal injury.

Combat Casualty Care and Preventive Medicine

4. Straighten his legs. Take the arm nearest to you, and move it so that it is straight and above his head. Repeat for the other arm.
5. Kneel beside the casualty with your knees near his shoulders. Leave room to roll his body (B, Figure 3-1). Place one hand behind his head and neck for support. With your other hand, grasp him under his far arm (C, Figure 3-1).
6. Roll him towards you with a steady, even pull. Keep his head and neck in line with his back.
7. Return his arms to his side. Straighten his legs, and reposition yourself so that you are kneeling at the level of his shoulders.
8. If you suspect a neck injury, and you are planning to use the jaw-thrust technique, then kneel at the casualty's head while looking towards his feet.

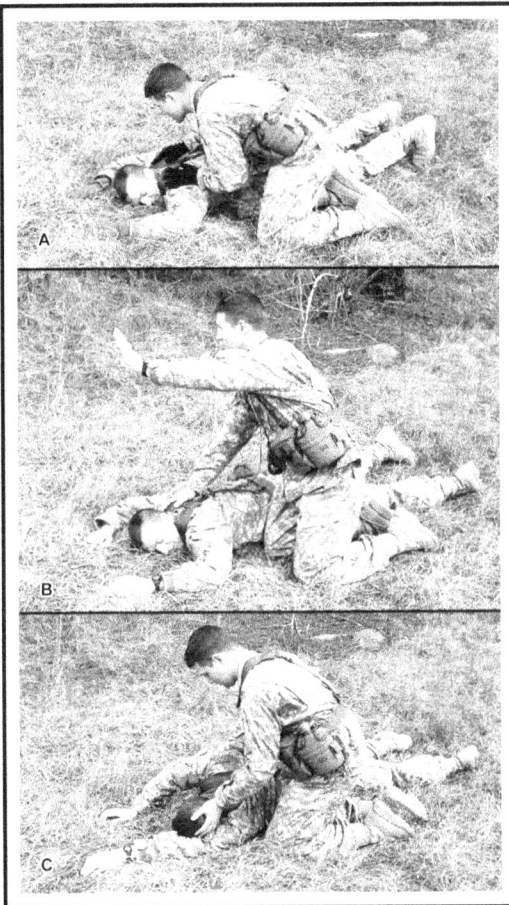

Figure 3-1. Assessment.

Chapter 3

Open Airway of Unconscious or Nonbreathing Casualty

3-9.   The tongue is the single most common airway obstruction (Figure 3-2). In most cases, just using the head-tilt/chin-lift technique can clear the airway. This pulls the tongue away from the air passage (Figure 3-3).

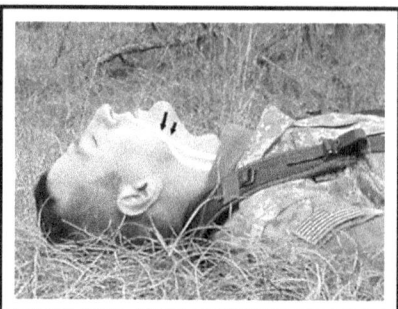

Figure 3-2. Airway blocked by tongue.

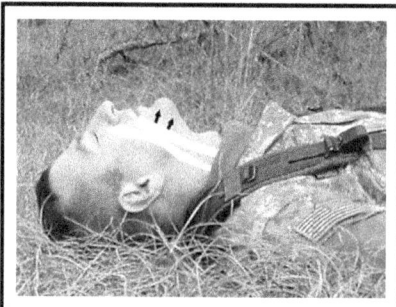

Figure 3-3. Airway opened by extending neck.

3-10.   Call for help, and then position the casualty. Move (roll) him onto his back (C, Figure 3-1). Perform a finger sweep. If you see foreign material or vomit in the casualty's mouth, promptly remove it, but avoid spending much time doing so. Open the airway using the jaw-thrust or head-tilt/chin-lift technique.

CAUTION

Although the head-tilt/chin-lift technique is an important procedure in opening the airway, take extreme care with it, because using too much force while performing this maneuver can cause more spinal injury. In a casualty with a suspected neck injury or severe head trauma, the safest approach to opening the airway is the jaw-thrust technique because, in most cases, you can do it without extending the casualty's neck.

Combat Casualty Care and Preventive Medicine

Perform Jaw-Thrust Technique

3-11. Place your hands on both sides of the angles of the casualty's lower jaw, and lift with both hands. Displace the jaw forward and up (Figure 3-4). Your elbows should rest on the surface where the casualty is lying. If his lips close, you can use your thumb to retract his lower lip. If you have to give mouth-to-mouth, then close his nostrils by placing your cheek tightly against them. Carefully support his head without tilting it backwards or turning it from side to side. This technique is the safest, and thus the first, to use to open the airway of a casualty who has a suspected neck injury. Why? Because, you can usually do it *without* extending his neck. However, if you are having a hard time keeping his head from moving, you might have to try tilting his head back *very* slightly.

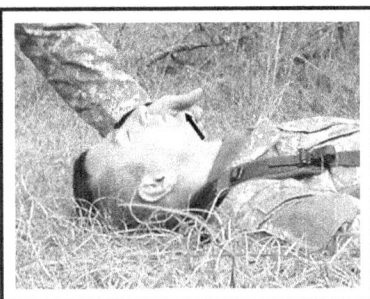

Figure 3-4. Jaw-thrust technique.

Perform Head-Tilt/Chin-Lift Technique

3-12. Place one palm on the casualty's forehead and apply firm, backward pressure to tilt his head back. Place the fingertips of your other hand under the bony part of his lower jaw, and then lift, bringing his chin forward. Avoid using your thumb to lift his chin (Figure 3-5).

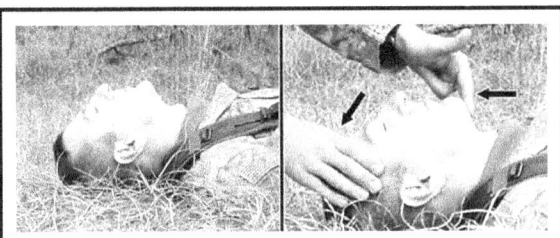

Figure 3-5. Head-tilt/chin-lift technique.

WARNING

Avoid pressing too deeply into the soft tissue under the casualty's chin, because you might obstruct his airway.

Chapter 3

Check for Breathing while Maintaining Airway

3-13. After opening the casualty's airway, you must keep it open. Often this is enough to let the casualty breathe properly. Failure to maintain the open airway will keep the casualty from receiving sufficient oxygen. While maintaining an open airway, check for breathing by observing the casualty's chest, and then, within a period of three to five seconds (Figure 3-6)--

1. *Look* for his chest to rise and fall.
2. *Listen* for sound of breathing by placing your ear near his mouth.
3. *Feel* for the flow of air on your cheek.
4. *Perform* rescue breathing if he fails to resume breathing spontaneously.

Note: If the casualty resumes breathing, monitor and maintain the open airway. Ensure he is transported to a medical treatment facility as soon as possible. Although the casualty might be trying to breathe, his airway might still be obstructed. If so, open his airway (remove the obstruction) and keep the airway open (maintain his airway).

Figure 3-6. Check for breathing.

Perform Rescue Breathing or Artificial Respiration

3-14. If the casualty fails to promptly resume adequate spontaneous breathing after the airway is open, you must start rescue breathing (artificial respiration, or mouth-to-mouth). Remain calm, but think and act quickly. The sooner you start rescue breathing, the more likely you are to restore his breathing. If you are not sure if the casualty is breathing, give him artificial respiration anyway. It cannot hurt him. If he is breathing, you can see and feel his chest move and, if you put your hand or ear close to his mouth and nose, you can hear him expelling air. The preferred method of rescue breathing is mouth-to-mouth, but you cannot always use it. For example, if the casualty has a severe jaw fracture or mouth wound, or if his jaws are tightly closed by spasms, you should use the mouth-to-nose method instead.

Combat Casualty Care and Preventive Medicine

Use Mouth-to-Mouth Method

3-15. In this best known method of rescue breathing, inflate the casualty's lungs with air from yours. You can do this by blowing air into his mouth. If the casualty is not breathing, place your hand on his forehead, and pinch his nostrils together with the thumb and index finger of the hand in use. With the same hand, exert pressure on his forehead to keep his head tilted backwards, and to maintain an open airway. With your other hand, keep your fingertips on the bony part of his lower jaw near his chin, and lift (Figure 3-5).

Note: If you suspect the casualty has a neck injury and you are using the jaw-thrust technique, close his nostrils by placing your cheek tightly against them.

3-16. Take a deep breath, and seal your mouth (airtight) around the casualty's mouth (Figure 3-7). If he is small, cover both his nose and mouth with your mouth, and then seal your lips against his face.

Figure 3-7. Rescue breathing.

3-17. Blow two full breaths into the casualty's mouth (1 to 1 1/2 seconds each), taking a fresh breath of air each time, before you blow. Watch from the corner of your eye for the casualty's chest to rise. If it does, then you are getting enough air into his lungs. If it fails to rise, then do the following:

1. Take corrective action immediately by reestablishing the airway. Ensure no air is leaking from around your mouth or from the casualty's pinched nose.
2. Try (again) to ventilate him.
3. If his chest still fails to rise, take the necessary action to open an obstructed airway.
4. If you are still unable to ventilate the casualty, reposition his head, and repeat rescue breathing. The main reason ventilation fails is improper chin and head positioning. If you cannot ventilate the casualty after you reposition his head, then move on to foreign-body airway obstruction maneuvers.
5. If, after you give two slow breaths, the casualty's chest rises, then see if you can find a pulse. Feel on the side of his neck closest to you by placing the index and middle fingers of your hand on the groove beside his Adam's apple (carotid pulse; Figure 3-8). Avoid using your thumb to take a pulse, because that could cause you to confuse your own pulse for his.

Chapter 3

6. Maintain the airway by keeping your other hand on the casualty's forehead. Allow 5 to 10 seconds to determine if there is a pulse.
7. If you see signs of circulation and you find a pulse, and the casualty has started breathing—
   a. *Stop* and allow the casualty to breathe on his own. If possible, keep him warm and comfortable.
   b. If you find a pulse, and the casualty is unable to breathe, continue rescue breathing until told to cease by medical personnel.
   c. If you fail to find a pulse, seek medical personnel for help as soon as possible.

Figure 3-8. Placement of fingers to detect pulse.

Use Mouth-to-Nose Method

3-18. Use this method if you cannot perform mouth-to-mouth rescue breathing. Normally, the reason you cannot is that the casualty has a severe jaw fracture or mouth wound, or, because his jaws are tightly closed by spasms. The mouth-to-nose method is the same as the mouth-to-mouth method, except that you blow into the *nose* while you hold the *lips* closed, keeping one hand at the chin. Then, you remove your mouth to let the casualty exhale passively. You might have to separate the casualty's lips to allow the air to escape during exhalation.

React to Airway Obstructions

3-19. For oxygen to flow to and from the lungs, the upper airway must be unobstructed. Upper airway obstruction can cause either partial or complete airway blockage. Upper airway obstructions often occur because--

1. The casualty's tongue falls back into his throat while he is unconscious.
2. His tongue falls back and obstructs the airway.
3. He was unable to swallow an obstruction.
4. He regurgitated the contents of his stomach, and they blocked his airway.
5. He has suffered blood clots due to head and facial injuries.

Combat Casualty Care and Preventive Medicine

> *Note:* For an injured or unconscious casualty, correctly position him, and then create and maintain an open airway.

Determine Degree of Obstruction

3-20. The airway may be partially or completely obstructed.

Partial

3-21. The person might still have an air exchange. If he has enough, then he can cough forcefully, even though he might wheeze between coughs. Instead of interfering, encourage him to cough up the object on his own. If he is not getting enough air, his coughing will be weak, and he might be making a high-pitched noise between coughs. He might also show signs of shock. Help him and treat him as though he had a complete obstruction.

Complete

3-22. A complete obstruction (no air exchange) is indicated if the casualty cannot speak, breathe, or cough at all. He might clutch his neck and move erratically. In an unconscious casualty, a complete obstruction is also indicated if, after opening his airway, you cannot ventilate him.

Open Obstructed Airway, Casualty Lying Down or Unresponsive

3-23. Sometimes you must expel an airway obstruction in a casualty who is lying down, who becomes unconscious, or who is found unconscious (cause unknown; Figure 3-9):

1. If a conscious casualty, who is choking, becomes unresponsive--
   a. Call for help.
   b. Open the airway.
   c. Perform a finger sweep.
   d. Try rescue breathing. If an airway blockage prevents this,
   e. Remove the airway obstruction.
2. If a casualty is unresponsive when you find him (cause unknown)--
   a. Assess or evaluate the situation.
   b. Call for help.
   c. Position the casualty on his back.
   d. Open the airway.
   e. Establish breathlessness.
   f. Try to perform rescue breathing. If still unable to ventilate the casualty,
   g. Perform six to ten manual (abdominal or chest) thrusts.

Chapter 3

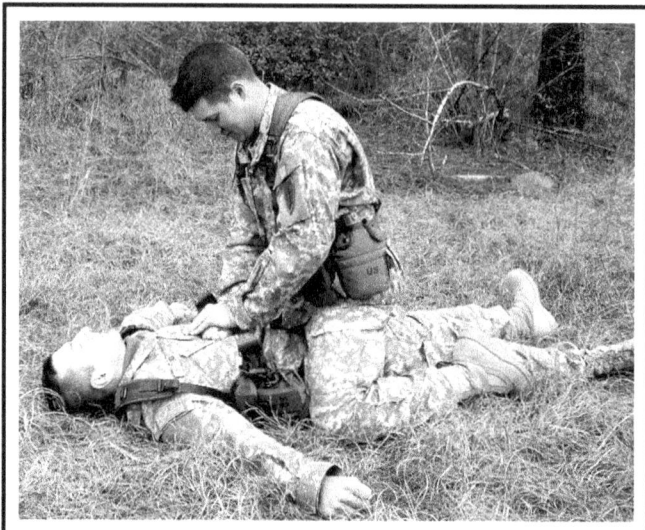

Figure 3-9. Abdominal thrust on unresponsive casualty.

3. To perform the abdominal thrusts--
   a. Kneel astride the casualty's thighs.
   b. Place the heel of one hand against the casualty's abdomen, in the midline slightly above the navel, but well below the tip of the breastbone.
   c. Place your other hand on top of the first one.
   d. Point your fingers toward the casualty's head.
   e. Use your body weight to press into the casualty's abdomen with a quick, forward and upward thrust.
   f. Deliver each thrust quickly and distinctly.
   g. Repeat the sequence of abdominal thrusts, finger sweep, and rescue breathing (try to ventilate) as long as necessary to remove the object from the obstructed airway.
   h. If the casualty's chest rises, check for a pulse.
4. To perform chest thrusts--
   a. Place the unresponsive casualty on his back, face up, and open his mouth.
   b. Kneel close to his side.
   c. Locate the lower edge of his ribs with your fingers.
   d. Run your fingers up along the rib cage to the notch (A, Figure 3-10).
   e. Place your middle finger on the notch, and your index finger next to your middle finger, on the lower edge of his breastbone.
   f. Place the heel of your other hand on the lower half of his breastbone, next to your two fingers (B, Figure 3-10).

3-10   TC 3-21.75   13 August 2013

Combat Casualty Care and Preventive Medicine

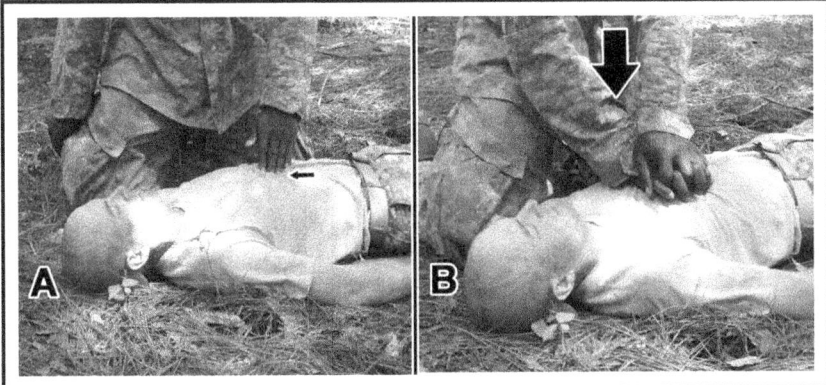

Figure 3-10. Hand placement for chest thrust.

g. Remove your fingers from the notch and place that hand on top of your hand on his breastbone, extending or interlocking your fingers.

h. Straighten and lock your elbows, with your shoulders directly above your hands. Be careful to avoid bending your elbows, rocking, or letting your shoulders sag. Apply enough pressure to depress the breastbone 1 1/2 to 2 inches, and then release the pressure completely. Repeat six to ten times. Deliver each thrust quickly and distinctly. Figure 3-11 shows another view of the breastbone being depressed.

Figure 3-11. Breastbone depressed 1 1/2 to 2 inches.

i. Repeat the sequence of chest thrust, finger sweep, and rescue breathing as long as necessary to clear the object from the obstructed airway.

j. If the casualty's chest rises, check his pulse.

5. If you still cannot administer rescue breathing due to an airway obstruction, remove the obstruction:
   a. Place the casualty on his back, face up.
   b. Turn him all at once (avoid twisting his body).
   c. Call for help.
   d. Perform finger sweep.
   e. Keep him face up.
   f. Use the tongue-jaw lift to open his mouth.
   g. Open his mouth by grasping both his tongue and lower jaw between your thumb and fingers, and lift (tongue-jaw lift; Figure 3-12).
   h. If you cannot open his mouth, cross your fingers and thumb (crossed-finger method), and push his teeth apart. To do this, press your thumb against his upper teeth, and your finger against his lower teeth (Figure 3-13).

Figure 3-12. Opening of casualty's mouth, tongue-jaw lift.

Figure 3-13. Opening of casualty's mouth, crossed-finger method.

   i. Insert the index finger of your other hand down along the inside of his cheek to the base of his tongue. Use a hooking motion from the side of the mouth toward the center to dislodge the foreign body (Figure 3-14).

Combat Casualty Care and Preventive Medicine

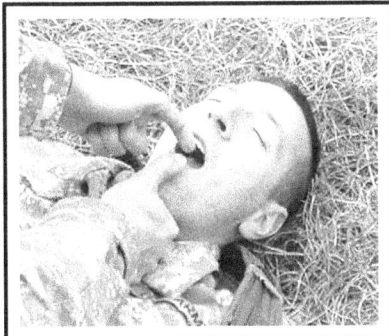

Figure 3-14. Use of finger to dislodge a foreign body.

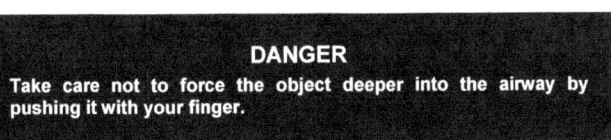

DANGER
Take care not to force the object deeper into the airway by pushing it with your finger.

CHECK FOR BLEEDING

Stop Bleeding and Protect Wound

3-24. The longer a Soldier bleeds from a major wound, the less likely he will survive it. (FM 4-25.11 covers first aid for open, abdominal, chest, and head wounds.) You must promptly stop the external bleeding.

Clothing

3-25. In evaluating him for location, type, and size of wound or injury, cut or tear the casualty's clothing and carefully expose the entire area of the wound. This is necessary to properly visualize the injury and avoid further contamination. To avoid further injury, leave in place any clothing that is stuck to the wound. *Do not* touch the wound, and keep it as clean as possible.

WARNING
In a chemical environment, leave a casualty's protective clothing in place. Apply dressings over the protective clothing.

Chapter 3

Entrance and Exit Wounds

3-26. Before applying the dressing, carefully examine the casualty to determine if there is more than one wound. A missile may have entered at one point and exited at another point. An exit wound is usually *larger* than its entrance wound.

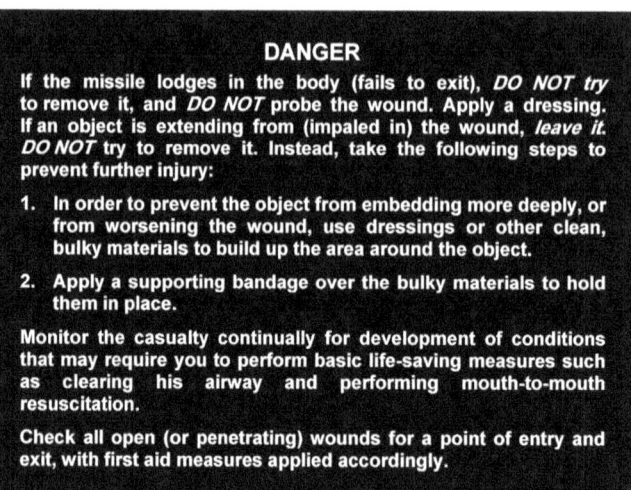

DANGER

If the missile lodges in the body (fails to exit), *DO NOT try* to remove it, and *DO NOT* probe the wound. Apply a dressing. If an object is extending from (impaled in) the wound, *leave it*. *DO NOT* try to remove it. Instead, take the following steps to prevent further injury:

1. In order to prevent the object from embedding more deeply, or from worsening the wound, use dressings or other clean, bulky materials to build up the area around the object.

2. Apply a supporting bandage over the bulky materials to hold them in place.

Monitor the casualty continually for development of conditions that may require you to perform basic life-saving measures such as clearing his airway and performing mouth-to-mouth resuscitation.

Check all open (or penetrating) wounds for a point of entry and exit, with first aid measures applied accordingly.

Emergency Trauma Dressing

3-27. Remove the emergency bandage from the wounded Soldier's pouch (Figure 3-15). (*Do not* use the one in your pouch.)

Figure 3-15. Emergency bandage.

Combat Casualty Care and Preventive Medicine

3-28. Place the pad on the wound, white side down, and wrap the elastic bandage around the injured limb or body part (A, Figure 3-16). Insert the elastic bandage into the pressure bar (B, Figure 3-17). Tighten the elastic bandage (C, Figure 3-18). Pull back, forcing the pressure bar down onto the pad (D, Figure 3-19). Wrap the elastic bandage tightly over the pressure bar, and wrap over all the edges of the pad (E, Figure 3-20). Secure the hooking ends of the closure bar into the elastic bandage. *Do not* create a tourniquet-like effect (F, Figure 3-21).

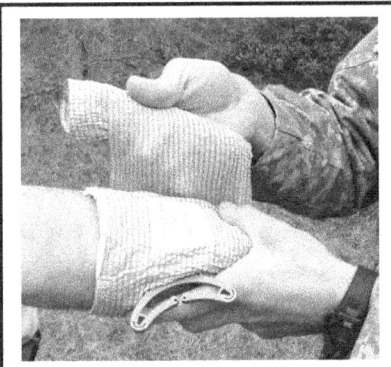

Figure 3-16. Application of pad to wound.

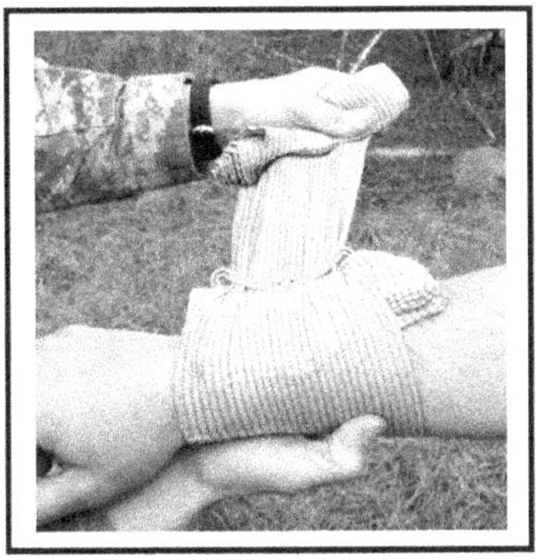

Figure 3-17. Insertion of bandage into pressure bar.

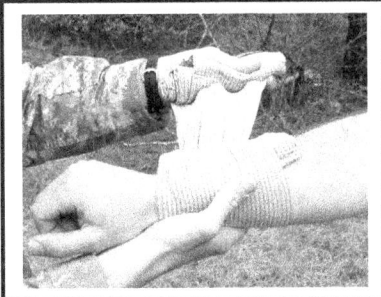

Figure 3-18. Tightening of bandage.

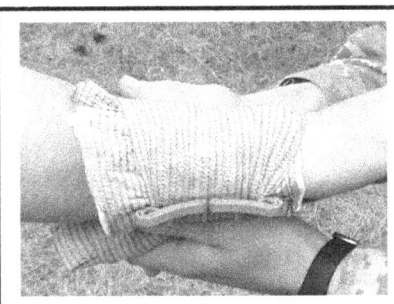

Figure 3-19. Pressure of bar into bandage.

Chapter 3

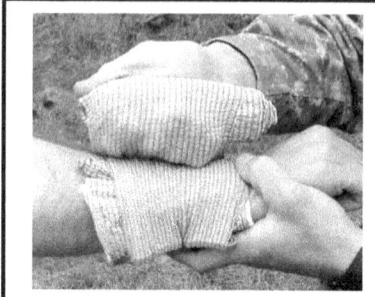 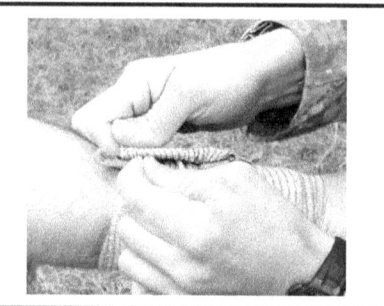

Figure 3-20.
Wrapping of bandage over pressure bar.

Figure 3-21.
Securing of bandage.

Field Dressing

3-29. Remove the casualty's field dressing from the wrapper, and grasp the tails of the dressing with both hands (Figure 3-22).

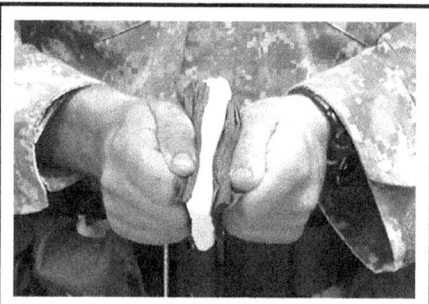

Figure 3-22. Grasping of dressing tails with both hands.

WARNING

Do not touch the white (sterile) side of the dressing.
Do not allow that side of the dressing to touch any surface other than the wound.

Combat Casualty Care and Preventive Medicine

3-30. Hold the dressing directly over the wound with the white side down. Open the dressing (Figure 3-23), and place it directly over the wound (Figure 3-24). Hold the dressing in place with one hand. Use the other hand to wrap one of the tails around the injured part, covering about half the dressing (Figure 3-25). Leave enough of the tail for a knot. If the casualty is able, he can help by holding the dressing in place.

Figure 3-23.
Pulling open of dressing.

Figure 3-24. Placement of dressing directly on wound.

Figure 3-25. Wrapping of dressing tail around injured part.

3-31. Wrap the other tail in the opposite direction until the rest of the dressing is covered. The tails should seal the sides of the dressing to keep foreign material from getting under it. Tie the tails into a nonslip knot over the outer edge of the dressing (Figure 3-26). *Do not tie the knot over the wound.* In order to allow blood to flow to the rest of the injured limb, tie the dressing firmly enough to prevent it from slipping, but without causing a tourniquet effect. That is, the skin beyond the injury should not become cool, blue, or numb.

Figure 3-26. Tails tied into nonslip knot.

Figure 3-27.
Application of direct manual pressure.

Chapter 3

Manual Pressure

3-32. If bleeding continues after you apply the sterile field dressing, apply direct pressure to the dressing for five to ten minutes (Figure 3-27). If the casualty is conscious and can follow instructions, you can ask him to do this himself. Elevate an injured limb slightly above the level of the heart to reduce the bleeding (Figure 3-28).

Figure 3-28. Elevation of injured limb.

WARNING

Elevate a suspected fractured limb *only after* properly splinting it.

3-33. If the bleeding stops, check for shock, and then give first aid for that as needed. If the bleeding continues, apply a pressure dressing.

Pressure Dressing

3-34. If bleeding continues after you apply a field dressing, direct pressure, and elevation, then you must apply a pressure dressing. This helps the blood clot, and it compresses the open blood vessel. Place a wad of padding on top of the field dressing directly over the wound (Figure 3-29). Keep the injured extremity elevated.

Note: Improvise bandages from strips of cloth such as tee shirts, socks, or other garments.

3-35. Place an improvised dressing (or cravat, if available) over the wad of padding (Figure 3-30). Wrap the ends tightly around the injured limb, covering the original field dressing (Figure 3-31).

Combat Casualty Care and Preventive Medicine

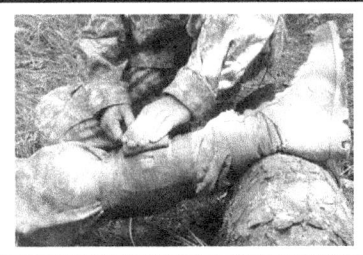

Figure 3-29. Wad of padding on top of field dressing.

Figure 3-30. Improvised dressing over wad of padding.

3-36. Tie the ends together in a nonslip knot, directly over the wound site (Figure 3-32). *Do not* tie so tightly that it has a tourniquet-like effect. If bleeding continues and all other measures fail, or if the limb is severed, then apply a tourniquet, but do so *only as a last resort*. When the bleeding stops, check for shock, and give first aid for that, if needed.

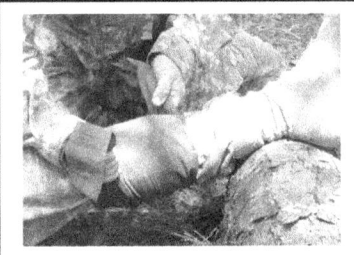

Figure 3-31. Ends of improvised dressing wrapped tightly around limb.

Figure 3-32. Ends of improvised dressing tied together in nonslip knot.

3-37. Check fingers and toes periodically for adequate circulation. Loosen the dressing if the extremity becomes cool, blue, or numb. If bleeding continues, and all other measures fail--application of dressings, covering of wound, direct manual pressure, elevation of limb above heart level, application of pressure dressing while maintaining limb elevation--then apply digital pressure.

Digital Pressure

3-38. Use this method when you are having a hard time controlling bleeding, before you apply a pressure dressing, or where pressure dressings are unavailable. Keep the limb elevated and direct pressure on the wound. At the same time, press your fingers, thumbs, or whole hand where a main artery supplying the wounded area lies near the surface or over bone (Figure 3-33). This might help shut off, or at least slow, the flow of blood from the heart to the wound.

Chapter 3

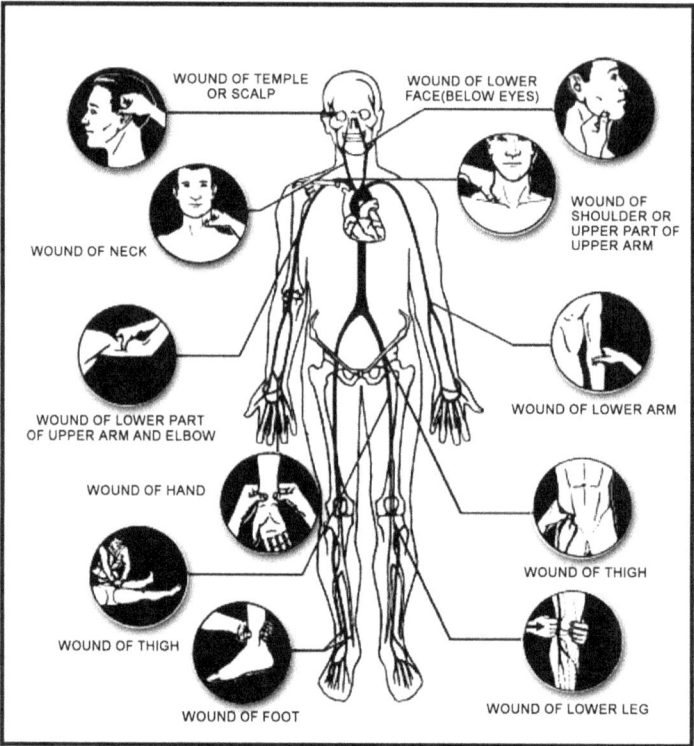

Figure 3-33. Digital pressure (fingers, thumbs, or hands).

Tourniquet

3-39.   A tourniquet is a constricting band placed around an arm or leg to control bleeding. A Soldier whose arm or leg has been completely amputated might not be bleeding when first discovered, but you should apply a tourniquet anyway. The body initially stops bleeding by contracting or clotting the blood vessels. However, when the vessels relax, or if a clot is knocked loose when the casualty is moved, the bleeding can restart. Bleeding from a major artery of the thigh, lower leg, or arm, and bleeding from multiple arteries, both of which occur in a traumatic amputation, might be more than you can control with manual pressure. If even under firm hand pressure the dressing gets soaked with blood, and if the wound continues to bleed, then you *must* apply a tourniquet.

DANGER
Use a tourniquet only on an arm or leg and if the casualty is in danger of bleeding to death.

WARNING
Continually monitor the casualty for the development of any conditions that could require basic life-saving measures such as clearing his airway, performing mouth-to-mouth breathing, preventing shock, or controlling bleeding.

Locate points of entry and exit on all open and penetrating wounds, and treat the casualty accordingly.

3-40. Avoid using a tourniquet unless a pressure dressing fails to stop the bleeding, or unless an arm or leg has been cut off. Tourniquets can injure blood vessels and nerves. Also, if left in place too long, a tourniquet can actually cause the loss of an arm or leg. However, that said, once you apply a tourniquet, you have to leave it in place and get the casualty to the nearest MTF ASAP. *Never* loosen or release a tourniquet yourself after you have applied one, because that could cause severe bleeding and lead to shock.

Combat Application Tourniquet

1. The C-A-T is packaged for one-handed use. Slide the wounded extremity through the loop of the C-A-T tape (1, Figure 3-34).
2. Position the C-A-T 2 inches above a bleeding site that is above the knee or elbow. Pull the free running end of the tape tight, and fasten it securely back on itself (2, Figure 3-34).
3. Do not affix the band past the windlass clip (3, Figure 3-34).
4. Twist the windlass rod until the bleeding stops (4, Figure 3-34).
5. Lock the rod in place with the windlass clip (5, Figure 3-34).
6. For small extremities, continue to wind the tape around the extremity and over the windlass rod (6, Figure 3-34).
7. Grasp the windlass strap, pull it tight, and adhere it to the hook-pile tape on the windlass clip (7, Figure 3-34). The C-A-T is now ready for transport.

Chapter 3

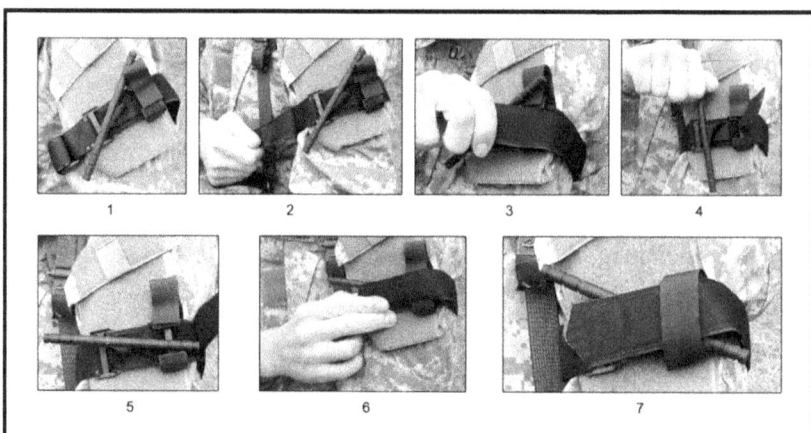

Figure 3-34. Band pulled tight.

> **WARNING**
>
> The one-handed method for upper extremities may not be completely effective on lower extremities.
>
> Ensure everyone receives familiarization and training on both methods of application.

3-41. The improved first-aid kit (IFAK) allows self-aid and buddy aid (SABA) interventions for extremity hemorrhages and airway compromises (Figure 3-35). The pouch and insert are both Class II items. Expendables are Class VIII.

Combat Casualty Care and Preventive Medicine

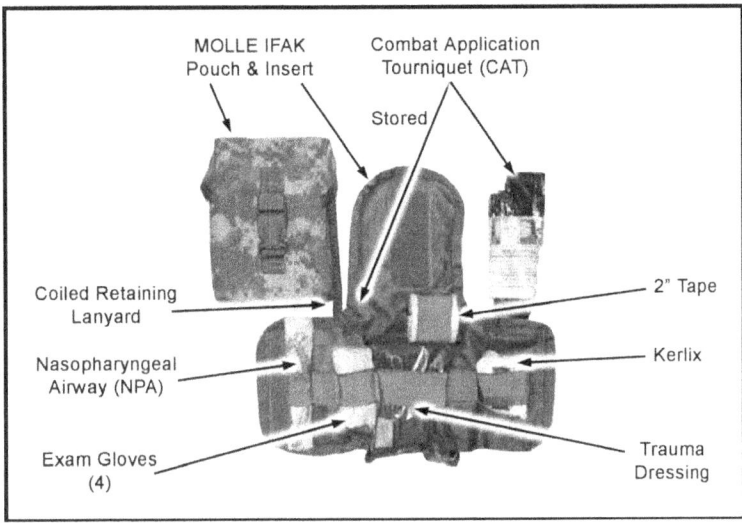

Figure 3-35. Improved first aid kit.

Improvised Tourniquet

3-42. In the absence of a specially designed tourniquet, you can make one from any strong, pliable material such as gauze or muslin bandages, clothing, or cravats. Use your improvised tourniquet with a rigid, stick-like object. To minimize skin damage, the improvised tourniquet must be at least 2 inches wide.

> **WARNING**
> The tourniquet must be easily identified or easily seen. Do not use wire, shoestring, or anything else that could cut into flesh, for a tourniquet band.

Placement

3-43. To position the makeshift tourniquet, place it around the limb, between the wound and the body trunk, or between the wound and the heart. *Never* place it directly over a wound, a fracture, or joint. For maximum effectiveness, place it on the upper arm or above the knee on the thigh (Figure 3-36).

Chapter 3

Figure 3-36. Tourniquet above knee.

3-44. Pad the tourniquet well. If possible, place it over a smoothed sleeve or trouser leg to keep the skin from being pinched or twisted. If the tourniquet is long enough, wrap it around the limb several times, keeping the material as flat as possible. Damaging the skin may deprive the surgeon of skin required to cover an amputation. Protecting the skin also reduces the casualty's pain.

Application

3-45. To apply the tourniquet, tie a half knot, which is the same as the first part of tying a shoe lace. Place a stick, or other rigid object, on top of the half knot (Figure 3-37).

Figure 3-37. Rigid object on top of half knot.

Combat Casualty Care and Preventive Medicine

3-46. Tie a full-knot over the stick, and twist the stick until the tourniquet tightens around the limb or the bright red bleeding stops (Figure 3-38). In the case of amputation, dark oozing blood may continue for a short time. This is the blood trapped in the area between the wound and tourniquet.

Figure 3-38. Tourniquet knotted over rigid object and twisted.

3-47. To fasten the tourniquet to the limb, loop the free ends of the tourniquet over the ends of the stick. Bring the ends around the limb to keep the stick from loosening. Tie the ends together on the side of the limb (Figure 3-39).

Figure 3-39. Free ends tied on side of limb.

Chapter 3

3-48. You can use other means to secure the stick. Just make sure the material remains wound around the stick, and that no further injury is possible. If possible, save and transport any severed (amputated) limbs or body parts with (but out of sight of) the casualty. Never cover the tourniquet. Leave it in full view. If the limb is missing (total amputation), apply a dressing to the stump. All wounds should have a dressing to protect the wound from contamination. Mark the casualty's forehead with a "T" and the time to show that he has a tourniquet. If necessary, use the casualty's blood to make this mark. Check and treat for shock, and then seek medical aid.

CAUTION

Do not remove a tourniquet yourself. Only trained medical personnel may adjust or otherwise remove or release the tourniquet, and then only in the appropriate setting.

SHOCK

3-49. The term *shock* means various things. In medicine, it means a collapse of the body's cardiovascular system, including an inadequate supply of blood to the body's tissues. Shock stuns and weakens the body. When the normal blood flow in the body is upset, death can result. Early recognition and proper first aid may save the casualty's life.

Causes and Effects of Shock

3-50. The three basic effects of shock are--

- Heart is damaged and fails to pump.
- Blood loss (heavy bleeding) depletes fluids in vascular system.
- Blood vessels dilate (open wider), dropping blood pressure to dangerous level.

3-51. Shock might be caused by--

- Dehydration.
- Allergic reaction to foods, drugs, insect stings, and snakebites.
- Significant loss of blood.
- Reaction to sight of wound, blood, or other traumatic scene.
- Traumatic injuries.
  -- Burns.
  -- Gunshot or shrapnel wounds.
     Crush injuries.
  -- Blows to the body, which can break bones or damage internal organs.
  -- Head injuries.
  -- Penetrating wounds such as from knife, bayonet, or missile.

Signs and Symptoms of Shock

3-52. Examine the casualty to see if he has any of the following signs and symptoms:

- Sweaty but cool (clammy) skin.
- Weak and rapid pulse.
- (Too) rapid breathing.
- Pale or chalky skin tone.
- Cyanosis (blue) or blotchy skin, especially around the mouth and lips.

Combat Casualty Care and Preventive Medicine

- Restlessness or nervousness.
- Thirst.
- Significant loss of blood.
- Confusion or disorientation.
- Nausea, vomiting, or both.

First-Aid Measures for Shock

3-53. First-aid procedures for shock in the field are the same ones performed to prevent it. When treating a casualty, always assume the casualty is in shock, or will be shortly. Waiting until the signs of shock are visible could jeopardize the casualty's life.

Casualty Position

3-54. *Never* move the casualty, or his limbs, if you suspect he has fractures, and they have not yet been splinted. If you have cover and the situation permits, move the casualty to cover. Lay him on his back. A casualty in shock from a chest wound, or who is having trouble breathing, might breathe easier sitting up. If so, let him sit up, but monitor him carefully, in case his condition worsens. Elevate his feet higher than the level of his heart. Support his feet with a stable object, such as a field pack or rolled up clothing, to keep them from slipping off.

WARNINGS

1. Do not elevate legs if the casualty has an unsplinted broken leg, head injury, or abdominal injury.

2. Check casualty for leg fracture(s), and splint them, if needed, before you elevate his feet. For a casualty with an abdominal wound, place his knees in an upright (flexed) position.

3-55. Loosen clothing at the neck, waist, or wherever it might be binding.

CAUTION

Do not loosen or remove protective clothing in a chemical environment.

3-56. Prevent the casualty from chilling or overheating. The key is to maintain normal body temperature. In cold weather, place a blanket or like item over and under him to keep him warm and prevent chilling. However, if a tourniquet has been applied, leave it exposed (if possible). In hot weather, place the casualty in the shade and protect him from becoming chilled; however, avoid the excessive use of blankets or other coverings. Calm the casualty. Throughout the entire procedure of providing first aid for a casualty, you should reassure the casualty and keep him calm. This can be done by being authoritative (taking charge) and by showing self-confidence. Assure the casualty that you are there to help him. Seek medical aid.

Food and Drink

3-57. When providing first aid for shock, *never* give the casualty food or drink. If you must leave the casualty, or if he is unconscious, turn his head to the side to prevent him from choking if he vomits.

Chapter 3

Casualty Evaluation

3-58. Continue to evaluate the casualty until medical personnel arrives or the casualty is transported to an MTF.

CASUALTY EVACUATION

3-59. Medical evacuation of the sick and wounded (with en route medical care) is the responsibility of medical personnel who have been provided special training and equipment. Therefore, wait for some means of medical evacuation to be provided unless a good reason for you to transport a casualty arises. When the situation is urgent and you are unable to obtain medical assistance or know that no medical evacuation assets are available, you will have to transport the casualty. For this reason, you must know how to transport him without increasing the seriousness of his condition.

3-60. Transport by litter is safer and more comfortable for a casualty than manual carries. It is also easier for you as the bearer(s). However, manual transportation might be the only feasible method, due to the terrain or combat situation. You might have to do it to save a life. As soon as you can, transfer the casualty to a litter as soon as you find or can improvise one.

MANUAL CARRIES

3-61. When you carry a casualty manually, you must handle him carefully and correctly to prevent more serious or possibly fatal injuries. Situation permitting, organize the transport of the casualty, and avoid rushing. Perform each movement as deliberately *and gently* as possible. Avoid moving a casualty until the type and extent of his injuries are evaluated, and the required first aid administered. Sometimes, you will have to move the casualty immediately, for example, when he is trapped in a burning vehicle. Manual carries are tiring, and can increase the severity of the casualty's injury, but might be required to save his life. Two-man carries are preferred, because they provide more comfort to the casualty, are less likely to aggravate his injuries, and are less tiring for the bearers. How far you can carry a casualty depends on many factors, such as--

- Nature of the casualty's injuries.
- Your (the bearer's or bearers') strength and endurance.
- Weight of the casualty.
- Obstacles encountered during transport (natural or manmade).
- Type of terrain.

ONE-MAN CARRIES

3-62. Use these carries when only one bearer is available to transport the casualty:

Fireman's Carry

3-63. This is one of the easiest ways for one person to carry another. After an unconscious or disabled casualty has been properly positioned (rolled onto his abdomen), raise him from the ground, and then support him and place him in the carrying position (Figure 3-40). Here's what you do:

- A. Position the casualty by rolling him onto his abdomen and straddle him. Extend your hands under his chest and lock them together (A, Figure 3-40).
- B. Lift him to his knees as you move backward (B, Figure 3-40).
- C. Continue to move backward, straightening his legs and locking his knees (C, Figure 3-40).
- D. Walk forward, bringing him to a standing position. Tilt him slightly backward to keep his knees from buckling (D, Figure 3-40).

E. Keep supporting him with one arm, and then free your other arm, quickly grasp his wrist, and raise his arm high. Immediately pass your head under his raised arm, releasing the arm as you pass under it (E, Figure 3-40).
F. Move swiftly to face the casualty and secure your arms around his waist. Immediately place your foot between his feet, and spread them apart about 6 to 8 inches (F, Figure 3-40).
G. Grasp the casualty's wrist, and raise his arm high over your head (G, Figure 3-40).
H. Bend down and pull the casualty's arm over and down on your shoulder, bringing his body across your shoulders. At the same time, pass your arm between his legs (H, Figure 3-40).
I. Grasp the casualty's wrist with one hand, and place your other hand on your knee for support (I, Figure 3-40).
J. Rise with the casualty positioned correctly. Your other hand should be free (J, Figure 3-40).

Chapter 3

Figure 3-40. Fireman's carry.

Combat Casualty Care and Preventive Medicine

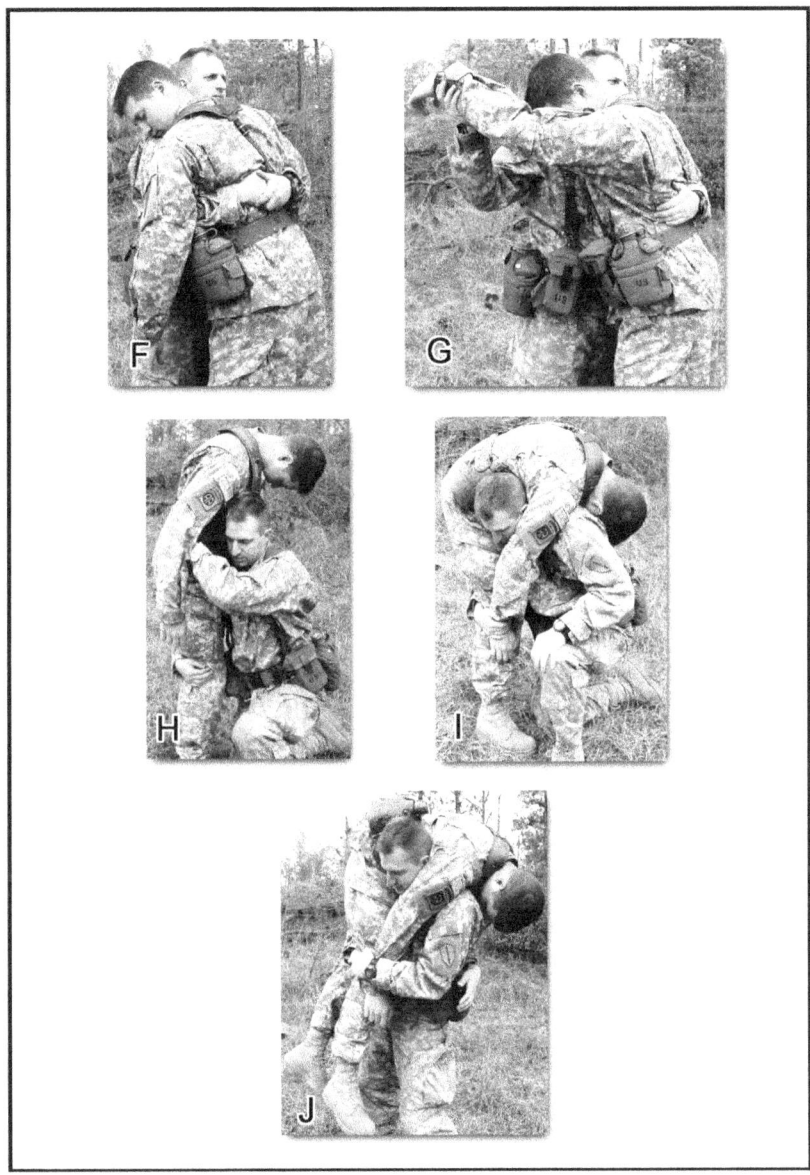

Figure 3-40. Fireman's carry (continued).

Chapter 3

Alternate Fireman's Carry

3-64.   Use this carry only when you think it is safer due to the location of the casualty's wounds. When you use the alternate carry, take care to keep the casualty's head from snapping back and injuring his neck. You can also use this method to raise a casualty from the ground for other one-man carries. First, kneel on both knees at the casualty's head and face his feet. Extend your hands under his armpits, down his sides, and across his back (A, Figure 3-41). Second, as you rise, lift the casualty to his knees. Then secure a lower hold and raise him to a standing position with his knees locked (B, Figure 3-41).

Figure 3-41. Alternate fireman's carry.

Supporting Carry

3-65.   With this method (Figure 3-42), the casualty must be able to walk or at least hop on one leg, with you as a crutch. You can use this carry to help him go as far as he can walk or hop. Raise him from the ground to a standing position using the fireman's carry. Grasp his wrist, and draw his arm around your neck. Place your arm around his waist. This should enable the casualty to walk or hop, with you as a support.

Combat Casualty Care and Preventive Medicine

Figure 3-42. Supporting carry.

Neck Drag

3-66. This method (Figure 3-43) is useful in combat, because you can carry the casualty as he creeps behind a low wall or shrubbery, under a vehicle, or through a culvert. If the casualty is conscious, let him clasp his hands together around your neck. To do this, first tie his hands together at the wrists, and then straddle him. You should be kneeling, facing the casualty. Second, loop his tied hands over and around your neck. Third, crawl forward and drag the casualty with you. If he is unconscious, protect his head from the ground.

WARNING

Avoid using this carry if the casualty has a broken arm.

Chapter 3

Figure 3-43. Neck drag.

Cradle-Drop Drag

3-67. Use this method to move a casualty up or down steps. Kneel at the casualty's head (with him on his back). Slide your hands, with palms up, under the casualty's shoulders. Get a firm hold under his armpits (A, Figure 3-44). Rise partially while supporting the casualty's head on one of your forearms (B, Figure 3-44). You may bring your elbows together and let the casualty's head rest on both of your forearms. Rise and drag the casualty backward so he is in a semi-seated position (C, Figure 3-44).

Combat Casualty Care and Preventive Medicine

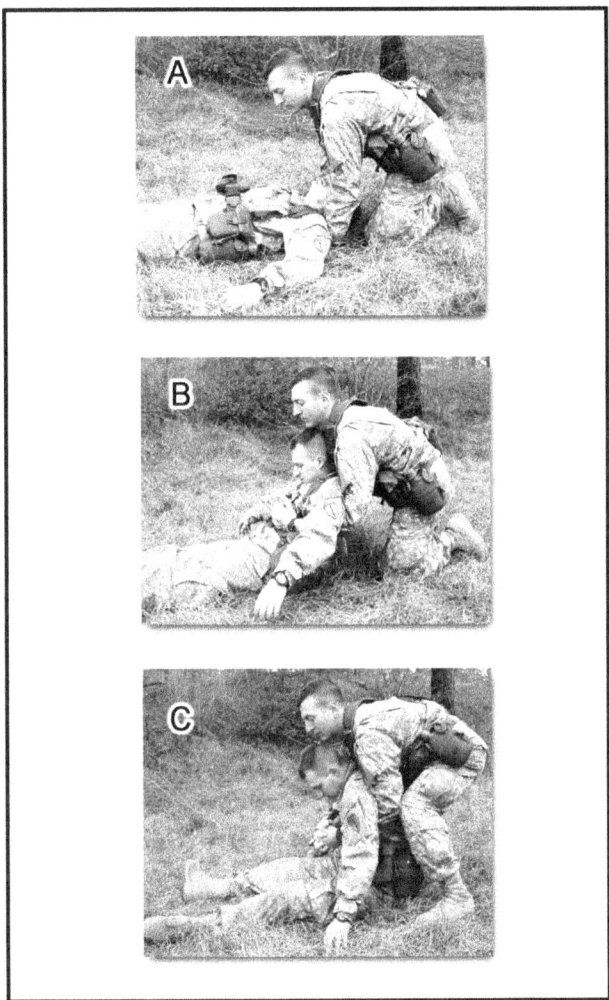

Figure 3-44. Cradle drop drag.

TWO-MAN CARRIES

3-68. Use these when you can. They are more comfortable to the casualty, less likely to aggravate his injuries, and less tiring for you.

Two-Man Support Carry

3-69. Use this method to transport either conscious or unconscious casualties. If the casualty is taller than you (the bearers), you might have to lift his legs and let them rest on your forearms. Help him to his feet, and then support him with your arms around his waist (A, Figure 3-45). Then, grasp the casualty's wrists and draw his arms around your necks (B, Figure 3-45).

Figure 3-45. Two-man support carry.

Two-Man Fore-and-Aft Carry

3-70. You can use this to transport a casualty for a long distance, say, over 300 meters. The taller of you (the two bearers) should position yourself at the casualty's head.

3-71. The shorter of you spreads the casualty's legs and kneels between them, with your back to the casualty. Position your hands behind the casualty's knees. The taller of you kneels at the casualty's head, slides your hands under his arms and across his chest, and locks your hands together (A, Figure 3-46). Both of you should rise together, lifting the casualty (B, Figure 3-46). If you alter this carry so that both of you are facing the casualty, you can use it to place him on a litter.

Combat Casualty Care and Preventive Medicine

Figure 3-46. Two-man fore-and-aft carry.

Two-Hand Seat Carry

3-72. You can use this method to carry a casualty for a short distance or to place him on a litter. With the casualty lying on his back (A, Figure 3-47), one of you should kneel on one side of the casualty at his hips, and the other should kneel on the other side. Each of you should pass your arms under the casualty's thighs and back, and grasp the other bearer's wrists. Both of you then rise, lifting the casualty (B, Figure 3-47).

Figure 3-47. Two-hand seat carry.

IMPROVISED LITTERS

3-73. Two men can support or carry a casualty without equipment for only short distances. By using available materials to improvise equipment, two or more rescuers can transport the casualty over greater distances.

1. Sometimes, a casualty must be moved without a standard litter. The distance might be too great for a manual carry, or the casualty might have an injury, such as a fractured neck, back, hip, or thigh, that manual transportation would aggravate. If this happens, improvise a litter from materials at hand. Construct it well to avoid dropping or further injuring the casualty. An improvised litter is an emergency measure only. Replace it with a standard litter as soon as you can.
2. You can improvise many types of litters, depending on the materials available. You can make a satisfactory litter by securing poles inside such items as ponchos, tarps, jackets, or shirts. You can improvise poles from strong branches, tent supports, skis, lengths of pipe, or other objects. If nothing is available to use as a pole, then roll a poncho or similar item from both sides toward the center, so you can grip the roll(s) and carry the casualty. You can use most any flat-surfaced object as long as it is the right size, for example, doors, boards, window shutters, benches, ladders, cots, or chairs. Try to find something to pad the litter for the casualty's comfort. You can use either the two-man fore-and-aft carry (Figure 3-46) or the two-hand seat carry (Figure 3-47) to place the casualty on a litter.
3. Use either two or four service members (head/foot) to lift a litter. Everybody should raise the litter at the same time to keep the casualty as level as possible.

> **DANGER**
> Unless there is an immediate life-threatening situation (such as fire or explosion), NEVER move a casualty who has a suspected back or neck injury. Instead, seek medical personnel for guidance on how to transport him.

> **WARNING**
> Use caution when transporting on a sloping incline/hill.

Section II. PREVENTIVE MEDICINE

Personal hygiene and cleanliness practices (Figure 3-48) safeguard your health and that of others. Specifically, they--

- Protect against disease-causing germs that are present in all environments.
- Keep disease-causing germs from spreading.
- Promote health among Soldiers.
- Improve morale.

- Never consume foods and beverages from unauthorized sources.
- Never soil the ground with urine or feces. Use a latrine or "cat hole."
- Keep your fingers and contaminated objects out of your mouth.
- Wash your hands--
  - After any contamination.
  - Before eating or preparing food.
  - Before cleaning your mouth and teeth.
- Wash all mess gear after each meal or use disposable plastic ware once.
- Clean your mouth and teeth at least once each day.
- Avoid insect bites by wearing proper clothing and using insect repellents.
- Avoid getting wet or chilled unnecessarily.
- Avoid sharing personal items with other Soldiers, for example--
  - Canteens.
  - Pipes.
  - Toothbrushes.
  - Washcloths.
  - Towels.
  - Shaving gear.
- Avoid leaving food scraps lying around.
- Sleep when possible.
- Exercise regularly.

Figure 3-48. Rules for avoiding illness in the field.

CLOTHING AND SLEEPING GEAR

3-74. Situation permitting, wash or exchange your clothing when it gets dirty. Do the same with your sleeping gear. When you cannot do this, at least shake everything out and air it regularly in the sun. This will reduce the number of germs on them.

CARE OF THE FEET

3-75. Wash and dry your feet at least daily. Use foot powder on your feet to help kill germs, reduce friction on the skin, and absorb perspiration. Change your socks daily. As soon as you can after you cross a wet area, dry your feet, put on foot powder, and change socks (Figure 3-49).

Figure 3-49. Care of the feet.

FOOD AND DRINK

3-76. For proper development, strength, and survival, your body requires proteins, fats, and carbohydrates. It also requires minerals, vitamins, and water. Issued rations have those essential food substances in the right amounts and proper balance. So, primarily eat those rations. When feasible, heat your meals. This will make them taste better and will reduce the energy required to digest them. Avoid overindulging in sweets, soft drinks, alcoholic beverages, and other non-issued rations. They have little nutritional value, and are often harmful. Eat food only from approved sources. Drink water only from approved sources, or treat it with water purification tablets. To do this--

1. Fill your canteen with water, keeping trash and other objects out.
2. Add one purification tablet to a quart of clear water or
3. Add two tablets to a quart of cloudy or very cold water.
4. In the absence of purification tablets, boil water for 5 minutes.
5. Replace the cap loosely.
6. Wait 5 minutes.
7. Shake the canteen well, and let some of the water to leak out.
8. Tighten the cap.
9. Wait 20 more minutes before drinking the water.

MENTAL HEALTH AND MORALE

3-77. To maintain mental health and self confidence--

MENTAL HYGIENE

3-78. The way you think affects the way you act. If you know your job, you will probably act quickly and effectively. If you are uncertain or doubtful of your ability to do your job, you may hesitate and make wrong decisions. Positive thinking is a necessity. You must enter combat with absolute confidence in your ability to do your job. Keep in mind that--

- Fear is a basic human emotion. It is mental and physical. In itself, fear is not shameful, if controlled. It can even help you, by making you more alert and more able to do your job. For example, a fear-induced adrenaline rush might help you respond and defend yourself or your comrades quickly during an unpredicted event or combat situation. Therefore, fear can help you--use it to your advantage.
- Avoid letting your imagination and fear run wild. Remember, you are not alone. You are part of a team. Other Soldiers are nearby, even though you cannot always see them. Everyone must help each other and depend on each other.
- Worry undermines the body, dulls the mind, and slows thinking and learning. It adds to confusion, magnifies troubles, and causes you to imagine things that really do not exist. If you are worried about something, talk to your leader about it. He might be able to help solve the problem.
- You might have to fight in any part of the world and in all types of terrain. Therefore, adjust your mind to accept conditions as they are. If mentally prepared for it, you should be able to fight under almost any conditions.

EXERCISE

3-79. Exercise your muscles and joints to maintain your physical fitness and good health. Without exercise, you might lack the physical stamina and ability to fight. Physical fitness includes a healthy body, the capacity for skillful and sustained performance, the ability to recover from exertion rapidly, the desire to complete a designated task, and the confidence to face any possible event. Your own safety, health, and life may depend on your physical fitness. During lulls in combat, counteract inactivity by exercising. This helps keep your muscles and body functions ready for the next period of combat. It also helps pass the time.

REST

3-80. Your body needs regular periods of rest to restore physical and mental vigor. When you are tired, your body functions are sluggish, and your ability to react is slower than normal, which makes you more susceptible to sickness, and to making errors that could endanger you or others. For the best health, you should get 6 to 8 hours of uninterrupted sleep each day. As that is seldom possible in combat, use rest periods and off-duty time to rest or sleep. Never be ashamed to say that you are tired or sleepy. However, *never* sleep on duty.

This page intentionally left blank.

Chapter 4
Environmental Conditions

In today's Army, Soldiers may deploy and fight anywhere in the world. They may go into the tropical heat of Central America, the deserts of the Middle East, and the frozen tundra of Alaska. Each environment presents unique situations concerning a Soldiers' performance. Furthermore, physical exertion in extreme environments can be life threatening. While recognizing such problems are important, preventing them is even more important; and furthermore, requires an understanding of the environmental factors that affect performance and how the body responds to those factors.

Section I. DESERT

Desert terrain, demanding and difficult to traverse, often provides very few landmarks. Furthermore, with cover and concealment highly limited, the threat of exposure to the enemy is constant. Most arid areas have several types of terrain.

TYPES

4-1. The five basic desert types are--

- Mountainous (high altitude).
- Rocky plateaus.
- Sand dunes.
- Salt marshes.
- Broken, dissected terrain (*gebels* or *wadis*).

MOUNTAINOUS DESERTS

4-2. Scattered ranges or areas of barren hills or mountains separated by dry, flat basins characterize mountainous deserts. High ground may rise gradually or abruptly from flat areas to several thousand meters above sea level. Most of the infrequent rainfall occurs on high ground and runs off rapidly in the form of flash floods. These floodwaters erode deep gullies and ravines, and deposit sand and gravel around the edges of the basins. Water rapidly evaporates, leaving the land as barren as before, although there may be short-lived vegetation. If enough water enters the basin to compensate for the rate of evaporation, shallow lakes may develop, such as the Great Salt Lake in Utah or the Dead Sea. Most of these lakes have a high salt content.

ROCKY PLATEAU DESERTS

4-3. Rocky plateau deserts have relatively slight relief interspersed with extensive flat areas with quantities of solid or broken rock at or near the surface. There may be steep-walled, eroded valleys, known as *wadis* in the Middle East and *arroyos* or canyons in the US and Mexico. Although their flat bottoms may be superficially attractive as assembly areas, the narrower valleys can be extremely dangerous to men and material due to flash flooding after rains. The Golan Heights is an example rocky plateau desert.

Chapter 4

SAND DUNES

4-4. Sand dune deserts are extensive flat areas covered with sand or gravel. "Flat" is a relative term, as some areas may contain sand dunes that are over 1,000 feet (300 meters) high and 10 to 15 miles (16 to 24 kilometers) long. Traffic ability in such terrain will depend on the windward or leeward slope of the dunes and the texture of the sand. However, other areas may be flat for 10,000 feet (3,000 meters) and more. Plant life may vary from none to scrub over 7 feet (2 meters) high. Examples of this type of desert include the edges of the Sahara, the empty quarter of the Arabian Desert, areas of California and New Mexico, and the Kalahari in South Africa.

SALT MARSHES

4-5. Salt marshes are flat, desolate areas sometimes studded with clumps of grass, but devoid of other vegetation. They occur in arid areas where rainwater has collected, evaporated, and left large deposits of alkali salts and water with a high salt concentration. The water is so salty it is undrinkable. A crust that may be 1 to 12 inches (2.5 to 30 centimeters) thick forms over the saltwater. Arid areas may contain salt marshes as many as hundreds of kilometers square. These areas usually support many insects, most of which bite. Avoid salt marshes, as this type of terrain is highly corrosive to boots, clothing, and skin. A good example salt marsh is the Shatt al Arab waterway along the Iran-Iraq border.

BROKEN AND DISSECTED TERRAIN

4-6. All arid areas contain broken or highly dissected terrain. Rainstorms that erode soft sand and carve out canyons form this terrain. A wadi may range from 10 feet (3 meters) wide and 7 feet (2 meters) deep to several hundred meters wide and deep. The direction a wadi takes varies as much as its width and depth. It twists and turns in a maze-like pattern. A wadi will give you good cover and concealment, but be cautious when deciding to try to move through it, because it is very difficult terrain to negotiate.

PREPARATION

4-7. Surviving in an arid area depends on what you know and how prepared you are for the environmental conditions.

FACTORS

4-8. In a desert area, you must consider--

- Low rainfall.
- Intense sunlight and heat.
- Wide temperature range.
- Sparse vegetation.
- High mineral content near ground surface.
- Sandstorms.
- Mirages.

Low Rainfall

4-9. Low rainfall is the most obvious environmental factor in an arid area. Some desert areas receive less than 4 inches (10 centimeters) of rain annually. When they do, it comes as brief torrents that quickly run off the ground surface.

Environmental Conditions

Intense Sunlight And Heat

4-10. Intense sunlight and heat are present in all arid areas. Air temperature can rise as high as 140° F (60° C) during the day. Heat gain results from direct sunlight, hot blowing sand-laden winds, reflective heat (the sun's rays bouncing off the sand), and conductive heat from direct contact with the desert sand and rock. The temperature of desert sand and rock typically range from 30 to 40° F (16 to 22° C) more than that of the air. For example, when the air temperature is 110° F (43° C), the sand temperature may be 140° F (60° C). Intense sunlight and heat increase the body's need for water. Radios and sensitive equipment items exposed to direct intense sunlight can malfunction.

Wide Temperature Range

4-11. Temperatures in arid areas may get as high as 130° F (55° C) during the day, and as low as 50° F (10° C) at night. The drop in temperature at night occurs rapidly and will chill a person who lacks warm clothing and is unable to move about. The cool evenings and nights are the best times to work or travel.

Sparse Vegetation

4-12. Vegetation is sparse in arid areas; therefore, you will have trouble finding shelter and camouflaging your movements. During daylight hours, large areas of terrain are easily visible. If traveling in hostile territory, follow the principles of desert camouflage:

- Hide or seek shelter in dry washes (*wadis*) with thick vegetation and cover from oblique observation.
- Use the shadows cast from brush, rocks, or outcroppings. The temperature in shaded areas will be 52 to 63° F (11 to 17° C) cooler than the air temperature.
- Cover objects that will reflect the light from the sun.

4-13. Before moving, survey the area for sites that provide cover and concealment. Keep in mind that it will be difficult to estimate distance. The emptiness of desert terrain causes most people to underestimate distance by a *factor of three*: what appears to be 1/2 mile (1 kilometer) away is really 1.75 miles (3 kilometers) away.

Sandstorms

4-14. Sandstorms (sand-laden winds) occur frequently in most deserts. The Seistan desert wind in Iran and Afghanistan blows constantly for up to 120 days. In Saudi Arabia, winds can reach 77 mph (128 kph) in early afternoon. Major sandstorms and dust storms occur at least once a week. The greatest danger is getting lost in a swirling wall of sand. Wear goggles and cover your mouth and nose with cloth. If natural shelter is unavailable, mark your direction of travel, lie down, and wait out the storm. Dust and wind-blown sand interfere with radio transmissions. Therefore, plan to use other means of signaling such as pyrotechnics, signal mirrors, or marker panels, whichever you have.

Mirages

4-15. Mirages are optical phenomena caused by the refraction of light through heated air rising from a sandy or stony surface. Mirages occur in the desert's interior about 6 miles (10 kilometers) from the coast. They make objects that are 1 mile (1.5 kilometers) or more away appear to move. This mirage effect makes it difficult for you to identify an object from a distance. It also blurs distant range contours so much that you feel surrounded by a sheet of water from which elevations stand out as "islands." The mirage effect makes it hard for a person to identify targets, estimate range, and see objects clearly. However, if you can get to high ground 10 feet [3 meters] or more above the desert floor), you can get above the superheated air close to the ground and overcome the mirage effect. Mirages make land navigation difficult, because they obscure natural features. You can survey the area at dawn, dusk, or by moonlight when there is little likelihood of mirage. Light levels in desert areas are more intense than in other geographic areas. Moonlit

Chapter 4

nights are usually clear, with excellent visibility, because daytime winds die down and haze and glare disappear. You can see lights, red flashlights, and blackout lights at great distances. Sound carries very far as well. Conversely, during nights with little moonlight, visibility is extremely poor. Traveling is extremely hazardous. You must avoid getting lost, falling into ravines, or stumbling into enemy positions. Movement during such a night is practical only if you have a means to determine direction and have spent the day resting; observing and memorizing the terrain; and selecting your route.

NEED FOR WATER

4-16. Since the early days of World War II, when the US Army was preparing to fight in North Africa, the subject of Soldier and water in the desert has generated considerable interest and confusion. At one time, the US Army thought it could condition men to do with less water by progressively reducing their water supplies during training. This practice of water discipline has caused hundreds of heat casualties. A key factor in desert survival is understanding the relationship between physical activity, air temperature, and water consumption. The body requires a certain amount of water for a certain level of activity at a certain temperature. For example, a person performing hard work in the sun at 109° F (43° C) requires 19 liters (5 gallons) of water daily. Lack of the required amount of water causes a rapid decline in an individual's ability to make decisions and to perform tasks efficiently. Your body's normal temperature is 98.6° F (36.9° C). Your body gets rid of excess heat (cools off) by sweating. The warmer your body becomes—whether caused by work, exercises, or air temperature—the more you sweat. The more you sweat the more moisture you lose. Sweating is the principal cause of water loss. If you stop sweating during periods of high-air temperature, heavy work, or exercise, you will quickly develop heat stroke and require immediate medical attention. Understanding how the air temperature and your physical activity affect your water requirements allows you to take measures to get the most from your water supply. These measures are--

- Find shade and get out of the sun!
- Place something between you and the hot ground.
- Limit your movements!
- Conserve your sweat. Wear your complete uniform to include T-shirt. Roll the sleeves down, cover your head, and protect your neck with a scarf or similar item. These steps will protect your body from hot-blowing winds and the direct rays of the sun. Your clothing will absorb your sweat, keeping it against your skin so that you gain its full cooling effect.

4-17. Thirst is not a reliable guide for your need for water. A person who uses thirst as a guide will drink only two thirds of his daily water requirement. Drinking water at regular intervals helps your body remain cool and decreases sweating. Even when your water supply is low, sipping water constantly will keep your body cooler and reduce water loss through sweating. Conserve your fluids by reducing activity during the heat of day if possible. To prevent this voluntary dehydration, use the following guide:

- Below 100° F (38° C), drink 0.5 liter of water every hour.
- Above 100° F (38° C), drink 1 liter of water every hour.

HAZARDS

4-18. Several hazards are unique to the desert environment. These include insects, snakes, thorny plants and cacti, contaminated water, sunburn, eye irritation, and climatic stress. Insects of almost every type abound in the desert. Man, as a source of water and food, attracts lice, mites, wasps, and flies. Insects are extremely unpleasant and may carry diseases. Old buildings, ruins, and caves are favorite habitats of spiders, scorpions, centipedes, lice, and mites. These areas provide protection from the elements and attract other wildlife. Therefore, take extra care when staying in these areas. Wear gloves at all times in the desert. Do not place your hands anywhere without first looking to see what is there. Visually inspect an area before sitting or lying down. When you get up, shake out and inspect your boots and clothing. All desert areas have snakes. They inhabit ruins, native villages, garbage dumps, caves, and natural rock outcroppings that offer shade. Never go barefoot or walk through these areas without carefully inspecting them for snakes. Pay attention to where you place your feet and hands. Most snakebites result from stepping on or handling snakes. Avoid them. Once you see a snake, give it a wide berth.

Section II. JUNGLE

The jungle comprises a substantial portion of the earth's land mass. Jungle environments consist of tall grasslands; mountains; swamps; blue and brown water; and single/double-canopy vegetation. Jungle environments are prominent in South America, Asia, and Africa. High temperatures, heavy rainfall, and oppressive humidity characterize equatorial and subtropical regions, except at high altitudes. At low altitudes, temperature variation is seldom less than 50° F (10° C) and is often more than 95° F (35° C). At altitudes over 4,921 feet (1,500 meters), ice often forms at night. The rain has a cooling effect, but stops when the temperature soars. Rainfall is heavy, often with thunder and lightning. Sudden rain beats on the tree canopy, turning trickles into raging torrents and causing rivers to rise. Just as suddenly, the rain stops. Violent storms may occur, usually toward the end of the summer months. The dry season has rain once a day and the monsoon has continuous rain. In Southeast Asia, winds from the Indian Ocean bring the monsoon but the area is dry when the wind blows from the landmass of China. Tropical day and night are of equal length. Darkness falls quickly and daybreak is just as sudden. Leaders must consider several jungle subtypes and other factors when performing duty and surviving in the jungle:

TYPES

4-19. There is no standard type of jungle. Jungle can consist of any combination of the following terrain subtypes:

- Rain forests.
- Secondary jungles.
- Semi-evergreen seasonal and monsoon forests.
- Scrub and thorn forests.
- Savannas.
- Saltwater swamps.
- Freshwater swamps.

TROPICAL RAIN FORESTS

4-20. The climate varies little in rain forests. You find these forests across the equator in the Amazon and Congo basins, parts of Indonesia, and several Pacific islands. Up to 144 inches (365.8 centimeters) of rain falls throughout the year. Temperatures range from about 90° F (32° C) in the day to 70° F (21° C) at night. There are five layers of vegetation in this jungle. Sometimes still untouched by humans, jungle trees rise from buttress roots to heights of 198 feet (60 meters). Below them, smaller trees produce a canopy so thick that little light reaches the jungle floor. Seedlings struggle to reach light, and masses of vines twine their way to the sun. Ferns, mosses, and herbaceous plants push through a thick carpet of leaves, and fungi adorn leaves and fallen trees. The darkness of the jungle floor limits growth, which aids in movement. Little undergrowth is present to hamper movement, but dense growth limits visibility to about 55 yards (50 meters). You can easily lose your sense of direction in a tropical rain forest, and aircraft have a hard time seeing you.

SECONDARY JUNGLES

4-21. Secondary jungle is very similar to rain forest. Prolific growth, where sunlight penetrates to the jungle floor, typifies this type of forest. Such growth happens mainly along riverbanks, on jungle fringes, and where Soldiers have cleared rain forested areas. When abandoned, tangled masses of vegetation quickly reclaim these cultivated areas. You can often find cultivated food plants among secondary jungles.

SEMI-EVERGREEN SEASONAL AND MONSOON FORESTS

4-22. The characteristics of the American and African semi-evergreen seasonal forests correspond with those of the Asian monsoon forests:

- Their trees fall into two stories of tree strata.
    -- Upper story 60 to 79 feet (18 to 24 meters)
    -- Lower story 23 to 43 feet (7 to 13 meters)
- The diameter of the trees averages 2 feet (0.5 meter).
- Their leaves fall during a seasonal drought.

4-23. Except for the sago, nipa, and coconut palms, the same edible plants grow in these areas as in the tropical rain forests. You find these forests in portions of Columbia and Venezuela and the Amazon basin in South America; in southeast coastal Kenya, Tanzania, and Mozambique in Africa; in northeastern India, much of Burma, Thailand, Indochina, Java, and parts of other Indonesian islands in Asia.

TROPICAL SCRUB AND THORN FORESTS

4-24. Tropical scrub and thorn forests exist on the West coast of Mexico, on the Yucatan peninsula, in Venezuela, and in Brazil; on the Northwest coast and central parts of Africa; and (in Asia) in Turkistan and India. Food plants are scarce during the dry season, and more abundant during the rainy season. The chief characteristics of tropical scrub and thorn forests include—

- They have a definite dry season.
- Trees are leafless during the dry season.
- Ground is bare, except for a few tufted plants in bunches
- Grasses are uncommon.
- Plants with thorns predominate.
- Fires occur frequently.

TROPICAL SAVANNAS

4-25. South American savannas occur in parts of Venezuela, Brazil, and Guiana. In Africa, they occur in the southern Sahara (North central Cameroon and Gabon, and Southern Sudan); Benin; Togo; most of Nigeria; the Northeastern Republic of Congo; Northern Uganda; Western Kenya; and parts of Malawi and Tanzania, Southern Zimbabwe, Mozambique, and Western Madagascar. A savanna generally--

- Exists in the tropical zones of South America and Africa.
- Looks like a broad, grassy meadow, with trees spaced at wide intervals.
- Has lots of red soil.
- Grows scattered, stunted, and gnarled trees (like apple trees) as well as palm trees.

SALTWATER SWAMPS

4-26. Saltwater swamps are common in coastal areas subject to tidal flooding. Mangrove trees thrive in these swamps, and can grow to 39 feet (12 meters). In saltwater swamps, visibility is poor, and movement is extremely difficult. Sometimes, raftable streams form channels, but foot travel is usually required. Saltwater swamps exist in West Africa, Madagascar, Malaysia, the Pacific islands, Central and South America, and at the mouth of the Ganges River in India. Swamps at the mouths of the Orinoco and Amazon Rivers, and of the rivers of Guyana, offer plenty of mud and trees, but little shade. Tides in saltwater swamps can vary as much as 3 feet (0.9 meter). Advice for this terrain is, try to avoid the leeches, the various insects, including no-see-ums, and crocodiles and caimans. If you can, avoid saltwater swamps. However, if they have suitable water channels, you might be able to traverse them by raft, canoe, or rubber boat.

FRESHWATER SWAMPS

4-27. Freshwater swamps exist in some low-lying inland areas. They have masses of thorny undergrowth, reeds, grasses, and occasional short palms. These all reduce visibility and make travel difficult. Freshwater swamps are dotted with large and small islands, allowing you to get out of the water. Wildlife is abundant in freshwater swamps.

PREPARATION

4-28. Success in the jungle depends on your level of applicable knowledge and preparation.

TRAVEL THROUGH JUNGLE AREAS

4-29. With practice, you can move through thick undergrowth and jungle efficiently. Always wear long sleeves to avoid cuts and scratches.

"Jungle Eye"

4-30. To move easily, you must develop a "jungle eye." That is, look through the natural breaks in foliage rather than at the foliage itself. Stoop down occasionally to look along the jungle floor.

Chapter 4

Game Trails

4-31. You may find game trails you can follow. Stay alert and move slowly and steadily through dense forest or jungle. Stop periodically to listen and reorient on your objective. Many jungle and forest animals follow game trails. These trails wind and cross, but frequently lead to water or clearings. Use these trails if they lead in your desired direction of travel. However, they may also be favorite enemy points for ambushes and booby traps.

Machete

4-32. Use a machete to cut through dense vegetation, but avoid cutting too much, or you will tire quickly. If using a machete, stroke upward when cutting vines to reduce noise, because sound carries long distances in the jungle.

Stick

4-33. Use a stick to part the vegetation and to help dislodge biting ants, spiders, or snakes. Never grasp brush or vines when climbing slopes, because they may have irritating spines, sharp thorns, biting insects, and snakes.

Power and Telephone Lines

4-34. In many countries, electric and telephone lines run for miles through sparsely inhabited areas. Usually, the right-of-way is clear enough to allow easy travel. When traveling along these lines, be careful as you approach a transformer and relay stations, because they may be guarded.

WATER PROCUREMENT

4-35. Although water is abundant in most tropical environments, you may have trouble finding it, and when you do, it may not be safe to drink. Vines, roots, palm trees, and condensation are just a few of the many sources of water. You can sometimes follow animals to water. Often you can get nearly clear water from muddy streams or lakes by digging a hole in sandy soil about 3 feet (1 meter) from the bank. Water will then seep into the hole. Remember, you must purify any water you get this way.

POISONOUS PLANTS

4-36. The proportion of poisonous plants in tropical regions is no greater than in any other area of the world. However, it may appear that most plants in the tropics are poisonous, due to preconceived notions and the density of plant growth in some tropical areas.

Section III. ARCTIC

Cold regions include arctic and subarctic areas, and areas immediately adjoining them. About 48 percent of the Northern hemisphere's total land mass is a cold region, due to the influence and range of air temperatures. Ocean currents affect cold weather and cause large areas normally included in the temperate zone to fall within the cold regions during winter periods. Elevation also has a marked effect on defining cold regions. Within the cold weather regions, you may face two types of cold weather environments—wet or dry. Knowing which environment your area of operation (AO) falls in will affect planning and execution of a cold weather operation.

TYPES

4-37. The two types of arctic climates are wet-cold and dry-cold.

WET-COLD WEATHER ENVIRONMENTS

4-38. Wet-cold weather conditions exist when the average temperature in a 24-hour period is 14° F (-10° C) *or above*. Characteristics of this condition include freezing temperatures at night and slightly warmer temperatures during the day. Although temperatures in a wet-cold environment are warmer than those in a dry-cold environment, the terrain is usually very sloppy due to slush and mud. Protect yourself from the wet ground, freezing rain, and wet snow.

DRY-COLD WEATHER ENVIRONMENTS

4-39. Dry-cold weather conditions exist when the average temperature in a 24-hour period remains *below* 14° F (-10° C). Even though these temperatures are much lower than normal, you can avoid freezing and thawing. In temperatures down to -76° F (-60° C), wear extra layers of inner clothing. Wind and low temperatures are an extremely hazardous combination.

PREPARATION

4-40. Success in the arctic begins with preparedness:

WIND CHILL

4-41. Wind chill increases the hazards in cold regions. It is the effect of moving air on exposed flesh. For example, with a 15 knot (27.8 kmph) wind and a temperature of -14° F (-10° C), the equivalent wind chill temperature is -9° F (-23 degrees C). Remember, even when no wind is blowing, your own movement, such as during skiing, running, creates "apparent" wind, will create the equivalent wind by skiing, running, being towed on skis behind a vehicle, or working around aircraft that produce windblasts.

TRAVEL

4-42. Soldiers will find it almost impossible to travel in deep snow without snowshoes or skis. Traveling by foot leaves a well-marked trail for pursuers to follow. If you must travel in deep snow, avoid snow-covered streams. The snow, which acts as an insulator, may have prevented ice from forming over the water. In hilly terrain, avoid areas where avalanches appear possible. On ridges, snow gathers on the lee side in overhanging piles called cornices. These often extend far out from the ridge and may break loose if stepped on.

Chapter 4

WATER

4-43. Many sources of water exist in the arctic and subarctic. Your location and the season of the year will determine where and how you obtain water. Water sources in arctic and subarctic regions are more sanitary than in other regions due to the climatic and environmental conditions. However, always purify water before drinking it. During the summer months, the best natural sources of water are freshwater lakes, streams, ponds, rivers, and springs. Water from ponds or lakes may be slightly stagnant but still usable. Running water in streams, rivers, and bubbling springs is usually fresh and suitable for drinking.

PART TWO

Soldier Combat Skills

Chapter 5

Cover, Concealment, and Camouflage

If the enemy can see you and you are within range of his weapon system, he can engage and possibly kill you. So, you must be concealed from enemy observation and have cover from enemy fire. When the terrain does not provide natural cover and concealment, you must prepare your cover and use natural and man-made materials to camouflage yourself, your equipment, and your position. This chapter provides guidance on the preparation and use of cover, concealment, and camouflage, except for fighting positions, which are covered in Chapter 6.

Section I. COVER

Cover, made of natural or man-made materials, gives protection from bullets, fragments of exploding rounds, flame, nuclear effects, biological and chemical agents, and enemy observation (Figure 5-1). (Chapter 6 discusses cover for fighting positions.)

Figure 5-1. Natural cover.

Chapter 5

NATURAL COVER

5-1. Natural cover includes logs, trees, stumps, rocks, and ravines; whereas, man-made cover includes fighting positions, trenches, walls, rubble, and craters. To get protection from enemy fire in the offense or when moving, use routes that put cover between you and the enemy. For example, use ravines, gullies, hills, wooded areas, walls, and any other cover that will keep the enemy from seeing and firing at you (Figure 5-2). Avoid open areas. Never skyline yourself on a hilltop or ridge. Any cover--even the smallest depression or fold in the ground--can help protect you from direct and indirect enemy fire.

Figure 5-2. Cover along a wall.

MAN-MADE COVER

5-2. Man-made cover includes fighting positions and protective equipment.

FIGHTING POSITION

5-3. See Chapter 6 for a detailed discussion of fighting positions (Figure 5-3).

Figure 5-3. Man-made cover.

PROTECTIVE EQUIPMENT

5-4. Man-made cover can also be an article of protective equipment that can be worn such as body armor and helmet (Figure 5-4). Body armor is protective equipment that works as a form of armor to minimize injury from fragmentation and bullets. The interceptor body armor (IBA) system has an outer tactical vest (OTV) which is lined with finely woven Kevlar that will stop a 9-mm round and other slower moving fragments. It also has removable neck, throat, shoulder, and groin protection. Two small-arms protective inserts may also be added to the front and back of the vest, with each plate designed to stop 7.62-mm rounds. The plates are constructed of boron carbide ceramic with a shield backing that breaks down projectiles and halts their momentum. The vest also meets stringent performance specifications related to flexibility and heat stress requirements. The advanced combat helmet (ACH) provides protection against fragmentation and bullets, as well as heat and flame in a balanced and stable configuration.

Figure 5-4. Body armor and helmet.

SIMPLIFIED COLLECTIVE PROTECTION EQUIPMENT

5-5. The M20 simplified collective protection equipment (SCPE) is an inflatable shelter that provides cover against chemical/biological warfare agents and radioactive particles (Figure 5-5). The SCPE provides a clean-air environment in a structure where you can perform your duties, without wearing individual protective equipment.

Figure 5-5. Protective cover against chemical/biological warfare agents.

Section II. CONCEALMENT

Concealment is anything that hides you from enemy observation (Figure 5-6). Concealment does not protect you from enemy fire. Do not think that you are protected from the enemy's fire just, because you are concealed. Concealment, like cover, can also be natural or Soldier made. (Chapter 6 discusses techniques for concealing fighting positions.)

Figure 5-6. Concealment.

NATURAL CONCEALMENT

5-6. Natural concealment includes bushes, grass, and shadows. If possible, natural concealment should not be disturbed. Man-made concealment includes Army combat uniforms (ACUs), camouflage nets, face paint, and natural materials that have been moved from their original location. Man-made concealment must blend into natural concealment provided by the terrain.

ACTIONS AS CONCEALMENT

5-7. Light, noise, and movement discipline, and the use of camouflage, contributes to concealment. Light discipline is controlling the use of lights at night by such things as not smoking in the open, not walking around with a flashlight on, and not using vehicle headlights. Noise discipline is taking action to deflect sounds generated by your unit (such as operating equipment) away from the enemy and, when possible, using methods to communicate that do not generate sounds (arm-and-hand signals). Movement discipline includes not moving about fighting positions unless necessary and not moving on routes that lack cover and concealment. In the defense, build a well-camouflaged fighting position and avoid moving about. In the offense, conceal yourself and your equipment with camouflage, and move in woods or on terrain that gives concealment. Darkness cannot hide you from enemy observation in either offense or defense situations. The enemy's night vision devices (NVD) and other detection means allow them to find you in both daylight and darkness.

Cover, Concealment, and Camouflage

Section III. CAMOUFLAGE

Camouflage is anything you use to keep yourself, your equipment, and your position from being identified. Both natural and man-made material can be used for camouflage. Change and improve your camouflage often. The time between changes and improvements depends on the weather and on the material used. Natural camouflage will often die, fade, or otherwise lose its effectiveness. Likewise, man-made camouflage may wear off or fade and, as a result, Soldiers, their equipment, and their positions may stand out from their surroundings. To make it difficult for the enemy to spot them, Soldiers should remember the following when using or wearing camouflage. (Chapter 6 discusses techniques for camouflaging fighting positions.):

MOVEMENT

5-8. Movement and activity draw attention. When you give arm-and-hand signals or walk about your position, your movement can be seen by the naked eye at long ranges. In the defense, stay low. Move only when necessary. In the offense, move only on covered and concealed routes.

POSITIONS

5-9. Avoid putting anything where the enemy expects to find it. Build positions on the side of a hill, away from road junctions or lone buildings, and in covered and concealed places. Avoid open areas.

OUTLINES AND SHADOWS

5-10. These can reveal your position or equipment to an air or ground observer. Break up outlines and shadows with camouflage. When moving, try to stay in the shadows.

SHINE

5-11. A shine will naturally attract the enemy's attention. In the dark, a burning cigarette or flashlight will give you away. In daylight, reflected light from any polished surface such as shiny mess gear, a worn helmet, a windshield, a watch crystal and band, or exposed skin will do it. Any light, or reflection of light, can help the enemy detect your position. To reduce shine, cover your skin with clothing and face paint. Dull equipment and vehicle surfaces with paint, mud, or other camouflaging material or substance.

> **WARNING**
>
> In a nuclear attack, darkly painted skin can absorb more thermal energy and may burn more readily than bare skin.

SHAPE

5-12. Certain shapes, such as a helmet or human being, are easily recognizable. Camouflage, conceal, and break up familiar shapes to make them blend in with their surroundings, but avoid overdoing it.

Chapter 5

COLORS

5-13. If your skin, uniform, or equipment colors stand out against the background, the enemy can obviously detect you more easily than he could otherwise. For example, ACUs stand out against a backdrop of snow-covered terrain. Once again, camouflage yourself and your equipment to blend with the surroundings (Figure 5-7).

Figure 5-7. Soldier in arctic camouflage.

DISPERSION

5-14. This means spreading Soldiers, vehicles, and equipment over a wide area. The enemy can detect a bunch of Soldiers more easily than they can detect a lone Soldier. Spread out. Unit SOP or unit leaders vary distances between you and your fellow Soldiers depending on the terrain, degree of visibility, and enemy situation.

PREPARATION

5-15. Before camouflaging, study the terrain and vegetation of the area in which you are operating. Next, pick and use the camouflage material that best blends with the area (Figures 5-8). When moving from one area to another, change camouflage as needed to blend with the surroundings. Take grass, leaves, brush, and other material from your location and apply it to your uniform and equipment, and put face paint on your skin.

Cover, Concealment, and Camouflage

Figure 5-8. Camouflaged Soldiers.

INDIVIDUAL TECHNIQUES

HELMET

5-16. Camouflage your helmet with the issue helmet cover or make a cover of cloth or burlap that is colored to blend with the terrain (Figure 5-9). Leaves, grass, or sticks can also be attached to the cover. Use camouflage bands, strings, burlap strips, or rubber bands to hold those in place. If you have no material for a helmet cover, disguise and dull helmet surface with irregular patterns of paint or mud.

Chapter 5

Figure 5-9. Camouflaged helmet.

UNIFORM

5-17. The ACU has a jacket, trousers, and patrol cap in a new universal camouflage pattern. However, it may be necessary to add more camouflage to make the uniform blend better with the surroundings. To do this, put mud on the uniform or attach leaves, grass, or small branches to it. Too much camouflage, however, may draw attention. When operating on snow-covered ground wear overwhites (if issued) to help blend with the snow. If overwhites are not issued, use white cloth, such as white bed sheets, to get the same effect.

SKIN

5-18. Exposed skin reflects light and may draw the enemy's attention. Even very dark skin, because of its natural oil, will reflect light. The advanced camouflage face paint in compact form comes both with and without insect repellent. The active ingredient of the repellant is N, N-diethyl-m-toluamide (commonly known as DEET). The camouflage face paint provides visual and near-IR camouflage protection. The version with DEET also repels insects for eight hours. Both are furnished in compact form, and contain a full-sized, unbreakable, stainless steel mirror. Both compacts contain five compartments of pigmented formulations (green, loam, sand, white, and black). The compacts provide sufficient material for 20 applications of green, loam, and sand, and 10 applications of black and white. The compact is suitable for multi-terrain environmental conditions from arctic to desert. Face paints with insect repellent are supplied in a tan colored compact, while the non-repellent face paints are furnished in an olive drab compact for quick identification (Figure 5-10). When applying camouflage to your skin, work with a buddy (in pairs) and help each other. Apply a two-color combination of camouflage pigment in an irregular pattern. Do not apply camouflage paint if there is a chance of frostbite. The pigment may prevent other Soldiers from recognizing the whitish discoloration, the first symptoms of the skin freezing.

Cover, Concealment, and Camouflage

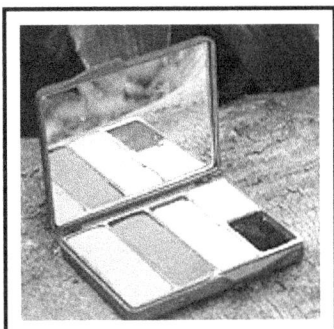

Figure 5-10. Advanced camouflage face paint.

Note: Advanced camouflage face paint with insect repellent, national stock number (NSN) 6840-01-493-7334, or without insect repellent, NSN 6850-01-493-7309.

5-19. Paint shiny areas (forehead, cheekbones, nose, ears, and chin) with a dark color. Paint shadow areas (around the eyes, under the nose, and under the chin) with a light color. In addition to the face, paint the exposed skin on the back of the neck, arms, and hands. Palms of hands are not normally camouflaged if arm-and-hand signals are to be used. Remove all jewelry to further reduce shine or reflection. When camouflage sticks/compacts are not issued, use burnt cork, bark, charcoal, lamp black, or light-colored mud (Table 5-1).

Table 5-1. Application of camouflage face paint to skin.

CAMOUFLAGE MATERIAL	SKIN COLOR	SHINE AREAS	SHADOW AREAS
	LIGHT OR DARK	FOREHEAD, CHEEKBONES, EARS, NOSE AND CHIN	AROUND EYES, UNDER NOSE, AND UNDER CHIN
LOAM AND LIGHT GREEN STICK	ALL TROOPS USE IN AREAS WITH GREEN VEGETATION	USE LOAM	USE LIGHT GREEN
SAND AND LIGHT GREEN STICK	ALL TROOPS USE IN AREAS LACKING GREEN VEGETATION	USE LIGHT GREEN	USE SAND
LOAM AND WHITE	ALL TROOPS USE ONLY IN SNOW-COVERED TERRAIN	USE LOAM	USE WHITE
BURNT CORK, BARK CHARCOAL, OR LAMP BLACK	ALL TROOPS, IF CAMOUFLAGE STICKS NOT AVAILABLE	USE	DO NOT USE
LIGHT-COLOR MUD	ALL TROOPS, IF CAMOUFLAGE STICKS NOT AVAILABLE	DO NOT USE	USE

This page intentionally left blank.

Chapter 6

Fighting Positions

Whether your unit is in a defensive perimeter or on an ambush line, you must seek cover from fire, and concealment from observation. From the time you prepare and occupy a fighting position, you should continue to improve it. How far you get depends on how much time you have, regardless of whether it is a hasty position or a well-prepared one with overhead cover (OHC). This chapter discusses--

- Cover and concealment.
- Sectors and fields of fire.
- Hasty and deliberate fighting positions.

COVER

6-1. To get this protection in the defense, build a fighting position to add to the natural cover afforded by the terrain (Figure 6-1). The cover of your fighting position will protect you from small arms fire and indirect fire fragments, and place a greater thickness of shielding material or earth between you and the blast wave of nuclear explosions.

Figure 6-1. Man-made cover.

6-2. Three different types of cover—overhead, frontal, and flank/rear cover—are used to make fighting positions. In addition, positions can be connected by tunnels and trenches. These allow Soldiers to move between positions for engagements or resupply, while remaining protected. (Chapter 5 discussed cover in general.)

Chapter 6

OVERHEAD COVER

6-3.   Your completed position should have OHC, which enhances survivability by protecting you from indirect fire and fragmentation.

FRONTAL COVER

6-4.   Your position needs frontal cover to protect you from small arms fire to the front. Frontal cover allows you to fire to the oblique, as well as to hide your muzzle flash.

FLANK AND REAR COVER

6-5.   When used with frontal and overhead cover, flank and rear cover protects you from direct enemy and friendly fire (Figure 6-2). Natural frontal cover such as rocks, trees, logs, and rubble is best, because it is hard for the enemy to detect. When natural cover is unavailable, use the dirt you remove to construct the fighting position. You can improve the effectiveness of dirt as a cover by putting it in sandbags. Fill them only three-quarters full.

Figure 6-2. Cover.

CONCEALMENT

6-6. If your position can be detected, it can be hit by enemy fire. Therefore, your position must be so well hidden that the enemy will have a hard time detecting it, even after he reaches hand-grenade range. (Chapter 5 discussed cover in general.)

NATURAL, UNDISTURBED MATERIALS

6-7. Natural, undisturbed concealment is better than man-made concealment. While digging your position, try not to disturb the natural concealment around it. Put the unused dirt from the hole behind the position and camouflage it. Camouflage material that does not have to be replaced (rocks, logs, live bushes, and grass) is best. Avoid using so much camouflage that your position looks different from its surroundings. Natural, undisturbed concealment materials--

- Are already prepared.
- Seldom attract enemy attention.
- Need no replacement.

MAN-MADE CONCEALMENT

6-8. Your position must be concealed from enemy aircraft as well as from ground troops. If the position is under a bush or tree, or in a building, it is less visible from above. Spread leaves, straw, or grass on the floor of the hole to keep freshly dug earth from contrasting with the ground around it. Man-made concealment must blend with its surroundings so that it cannot be detected, and must be replaced if it changes color or dries out.

CAMOUFLAGE

5-20. When building a fighting position, camouflage it and the dirt taken from it. Camouflage the dirt used as frontal, flank, rear, and overhead cover (OHC). Also, camouflage the bottom of the hole to prevent detection from the air. If necessary, take excess dirt away from the position (to the rear).

- Too much camouflage material may actually disclose a position. Get your camouflage material from a wide area. An area stripped of all or most of its vegetation may draw attention. Do not wait until the position is complete to camouflage it. Camouflage the position as you build.
- Hide mirrors, food containers, and white underwear and towels. Do not remove your shirt in the open. Your skin may shine and be seen. Never use fires where there is a chance that the flame will be seen or the smoke will be smelled by the enemy. Also, cover up tracks and other signs of movement. When camouflage is complete, inspect the position from the enemy's side. This should be done from about 38 feet (35 meters) forward of the position. Then check the camouflage periodically to ensure it is natural-looking and conceals the position. When the camouflage no longer works, change and improve it.

SECTORS AND FIELDS OF FIRE

6-9. Although a fighting position should provide maximum protection for you and your equipment, the primary consideration is always given to sectors of fire and effective weapons employment. Weapons systems are sited where natural or existing positions are available, or where terrain will provide the most protection while maintaining the ability to engage the enemy. You should always consider how best to use available terrain, and how you can modify it to provide the best sectors of fire, while maximizing the capabilities of your weapon system.

Chapter 6

SECTOR OF FIRE

6-10. A sector of fire is the area into which you must observe and fire. When your leader assigns you a fighting position, he should also assign you a primary and secondary sector of fire. The primary sector of fire is to the oblique of your position, and the secondary sector of fire is to the front.

FIELD OF FIRE

6-11. To be able to see and fire into your sectors of fire, you might have to "clear a field" of vegetation and other obstructions. Fields of fire are within the range of your weapons. A field of fire to the oblique lets you hit the attackers from an unexpected angle. It also lets you support the positions next to you. When you fire to the oblique, your fire interlocks with that of other positions, creating a wall of fire that the enemy must pass through. When clearing a field of fire--

- Avoid disclosing your position by careless or excessive clearing.
- Leave a thin, natural screen of vegetation to hide your position.
- In sparsely wooded areas, cut off lower branches of large, scattered trees.
- Clear underbrush only where it blocks your view.
- Remove cut brush, limbs, and weeds so the enemy will not spot them.
- Cover cuts on trees and bushes forward of your position with mud, dirt, or snow.
- Leave no trails as clues for the enemy.

HASTY AND DELIBERATE FIGHTING POSITIONS

6-12. The two types of fighting position are hasty and deliberate. Which you construct depends on time and equipment available, and the required level of protection. Fighting positions are designed and constructed to protect you and your weapon system. Table 6-1 shows the characteristics and planning considerations for fighting positions.

Fighting Positions

Table 6-1. Characteristics of individual fighting positions.

Type	Position	Estimated Construction Time (man-hours)	Equipment Requirements	Direct Small Caliber Fire	Indirect Fire Blast and Fragmentation (Near-Miss)*	Indirect Fire Blast and Fragmentation (Direct Hit)	Nuclear Weapons**	Remarks
Hasty	Crater	0.2	Hand tools	7.62 mm	Better than in open – no overhead protection	None	Fair	
Hasty	Skirmisher's trench	0.5	Hand tools	7.62 mm	Better than in open – no overhead protection	None	Fair	
Hasty	Prone position	1.0	Hand tools	7.62 mm	Better than in open – no overhead protection	None	Fair	Provides all-around cover
Deliberate	One-soldier position	3.0	Hand tools	12.7 mm	Medium artillery no closer than 30 ft – no overhead protection	None	Fair	
Deliberate	One-soldier position with 1.5 ft overhead cover	8.0	Hand tools	12.7 mm	Medium artillery no closer than 30 ft	None	Good	Additional cover provides protection from direct hit small mortar blast
Deliberate	Two-soldier position	6.0	Hand tools	12.7 mm	Medium artillery no closer than 30 ft – no overhead protection	None	Fair	
Deliberate	Two-soldier position with 1.5 ft overhead cover	11.0	Hand tools	12.7 mm	Medium artillery no closer than 30 ft	None	Good	Additional cover provides protection from direct hit small mortar blast
Deliberate	AT-4 position	3.0	Hand tools	12.7 mm	Medium artillery no closer than 30 ft – no overhead protection	None	Fair	

Note: Chemical protection is assumed because of individual protective masks and clothing.
* Shell sizes are: Small   Medium
   Mortar: 82mm   120mm
   Artillery: 105mm   152mm
** Nuclear protection ratings are rated poor, fair, good, very good, and excellent.

HASTY FIGHTING POSITION

6-13. Hasty fighting positions, used when there is little time for preparation, should be behind whatever cover is available. However, the term hasty does not mean that there is no digging. If a natural hole or ditch is available, use it. This position should give frontal cover from enemy direct fire but allow firing to the front and the oblique. When there is little or no natural cover, hasty positions provide as much protection as possible. A shell crater, which is 2 to 3 feet (0.61 to 1 meter) wide, offers immediate cover (except for overhead) and concealment. Digging a steep face on the side toward the enemy creates a hasty fighting position. A small crater position in a suitable location can later develop into a deliberate position. A skirmisher's trench is a shallow position that provides a hasty prone fighting position. When you need immediate shelter from enemy fire, and there are no defilade firing positions available, lie prone or on your side, scrape the soil with an entrenching tool, and pile the soil in a low parapet between yourself and the enemy. In all but the hardest ground, you can use this technique to quickly form a shallow, body-length pit. Orient the trench so it is oblique to enemy fire. This keeps your silhouette low, and offers some protection from small-caliber fire.

6-14. The prone position is a further refinement of the skirmisher's trench. It serves as a good firing position and provides you with better protection against the direct fire weapons than the crater position or the skirmisher's trench. Hasty positions are further developed into deliberate positions that provide as much protection as possible. The hole should be about 18 inches (46 centimeters) deep and use the dirt from the hole to build cover around the edge of the position (Figure 6-3).

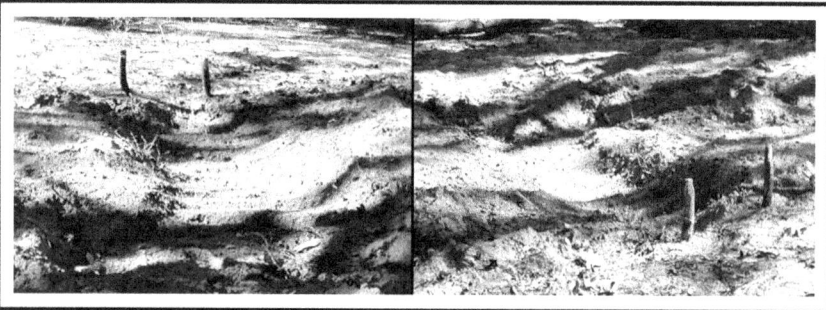

Figure 6-3. Prone position (hasty).

DELIBERATE FIGHTING POSITION

6-15. Deliberate fighting positions are modified hasty positions prepared during periods of relaxed enemy pressure. Your leader will assign the sectors of fire for your position's weapon system before preparation begins. Small holes are dug for automatic rifle bipod legs, so the rifle is as close to ground level as possible. Continued improvements are made to strengthen the position during the period of occupation. Improvements include adding OHC, digging grenade sumps, adding trenches to adjacent positions, and maintaining camouflage.

TWO-MAN FIGHTING POSITION

6-16. Prepare a two-man position in four stages. Your leader must inspect the position at each stage before you may move to the next stage (Table 6-2).

Fighting Positions

Table 6-2. Construction of two-man fighting position.

Parapets	Overhead Cover

Enable you to engage the enemy within your assigned sector of fire.
Provide you with protection from direct fire.
Construct parapets--
Thickness:
  Minimum 39 in (1m, length of M16 rifle)
Height:
  10 to 12 inches (25 to 30 centimeters, length of a bayonet) to the front, flank, and rear.
	Protect you from indirect fires. Your leaders will identify requirements for additional OHC based on threat capabilities.
	Thickness:
	  Minimum 18 inches (46 cm) (length of open entrenching tool)
	Concealment:
	  Use enough to make your position undetectable.

Note: If assigned an M4 rather than an M16-series weapon, add 7 inches (18 centimeters) to each dimension, on all positions that refer to the M16, or to two and a half M4 lengths.

Overhead Cover

6-17. Overhead cover may be built up or down.

Built-Up Overhead Cover

6-18. Built-up OHC has cover that is built up to 18 inches (46 centimeters) to maximize protection/cover of the fighting position.

Stage 1

6-19. Establish sectors and decide whether to build OHC up or down. Your leaders must consider the factors of mission, enemy, terrain, troops and equipment, time available, and civil considerations (METT-TC) in order to make a decision on the most appropriate fighting position to construct. For example due to more open terrain your leader may decide to use built-down OHC (Figure 6-4 and Figure 6-5):

1. Check fields of fire from the prone position.
2. Assign sector of fire (primary and secondary).
3. Emplace sector stakes (right and left) to define your sectors of fire. Sector stakes prevent accidental firing into friendly positions. Items such as tent poles, metal pickets, wooden stakes, tree branches, or sandbags will all make good sector stakes. The sector stakes must be sturdy and stick out of the ground at least 18 inches (46 centimeters); this will prevent your weapon from being pointed out of your sector.
4. Emplace aiming and limiting stakes to help you fire into dangerous approaches at night and at other times when visibility is poor. Forked tree limbs about 12 inches (30 centimeters) long make good stakes. Put one stake (possibly sandbags) near the edge of the hole to rest the stock of your rifle on. Then put another stake forward of the rear (first) stake/sandbag toward each dangerous approach. The forward stakes are used to hold the rifle barrel.
5. Emplace grazing fire logs or sandbags to achieve grazing fire 1 meter above ground level.
6. Decide whether to build OHC up or down, based on potential enemy observation of position.
7. Scoop out elbow holes to keep your elbows from moving around when you fire.
8. Trace position outline.

Chapter 6

9. Clear primary and secondary fields of fire.

Note: Keep in mind that the widths of all the fighting positions are only an approximate distance. This is due to the individual Soldier's equipment such as the IBA and the modular lightweight load-carrying equipment.

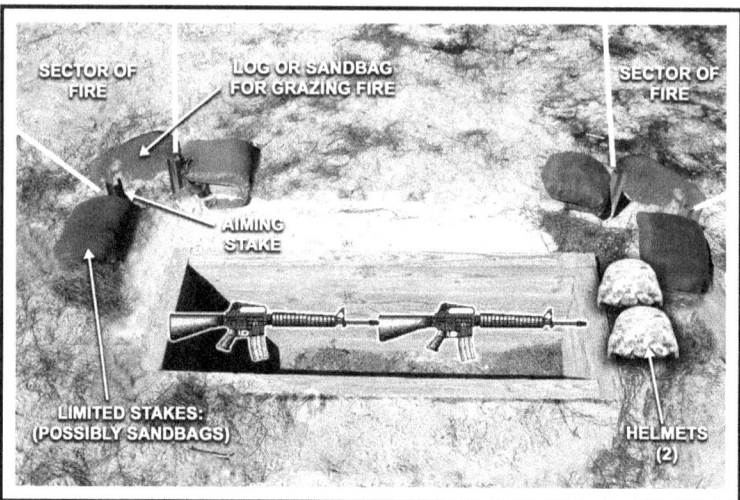

Figure 6-4. Establishment of sectors and building method.

Fighting Positions

Figure 6-5. Two-man fighting position (Stage 1).

Stage 2

6-20.   Place supports for OHC stringers and construct parapet retaining walls (Figure 6-6 and Figure 6-7):

1. Emplace OHC supports to front and rear of position.
2. Ensure you have at least 12 inches (30 centimeters), which is about 1-helmet length distance from the edge of the hole to the beginning of the supports needed for the OHC.
3. If you plan to use logs or cut timber, secure them in place with strong stakes from 2 to 3 inches (5 to 7 centimeters) in diameter and 18 inches (46 centimeters) long. Short U-shaped pickets will work.
4. Dig in about half the height.
    a. Front retaining wall--At least 10 inches (25 centimeters) high. (two filled sandbags) deep, and two M16s long.
    b. Rear retaining wall--At least 10 inches (25 centimeters) high, and one M16 long.
    c. Flank retaining walls--At least 10 inches (25 centimeters) high, and one M16 long.
5. Start digging hole; use soil to fill sandbags for walls.

Chapter 6

Figure 6-6. Placement of OHC supports and construction of retaining walls.

Figure 6-7. Two-man fighting position (Stage 2).

Fighting Positions

Stage 3

6-21. Dig position and place stringers for OHC (Figure 6-8, Figure 6-9, and Figure 6-10):

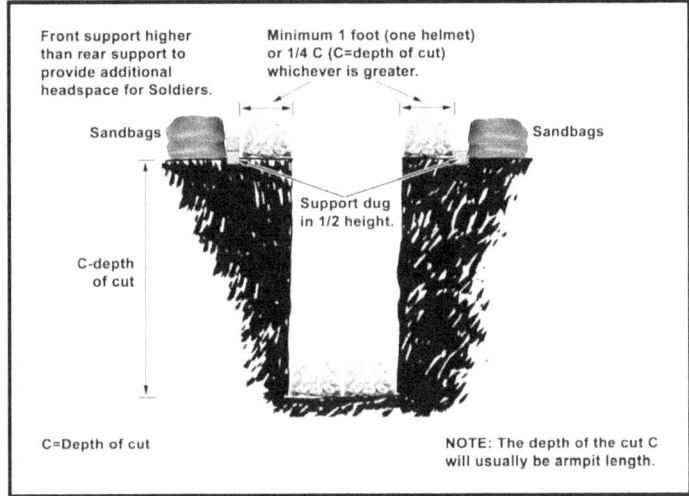

Figure 6-8. Digging of position (side view).

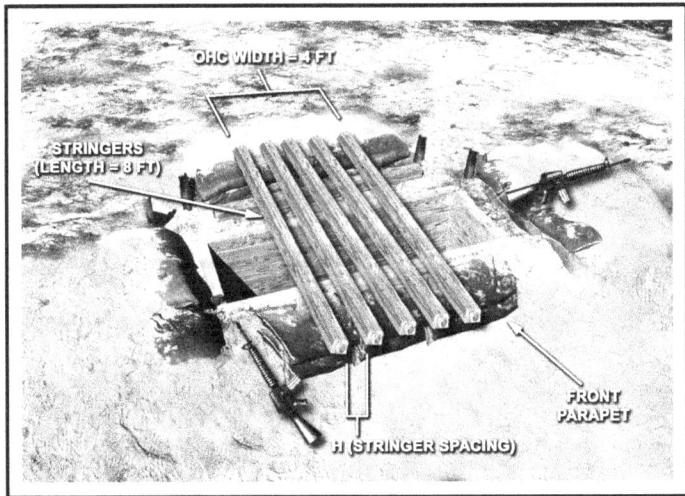

Figure 6-9. Placement of stringers for OHC.

Chapter 6

Figure 6-10. Two-man fighting position (Stage 3).

1. Ensure maximum depth is armpit deep (if soil conditions permit).
2. Use spoil from hole to fill parapets in the order of front, flanks, and rear.
3. Dig walls vertically.
4. If site soil properties cause unstable soil conditions, construct revetments (Figure 6-11) and consider sloping walls.

Figure 6-11. Revetment construction.

Fighting Positions

5. For sloped walls, first dig a vertical hole, and then slope walls at 1::4 ratio (move 12 inches [30 centimeters] horizontally for each 4 feet [1.22 meters] vertically).
6. Dig two grenade sumps in the floor (one on each end). If the enemy throws a grenade into the hole, kick or throw it into one of the sumps. The sump will absorb most of the blast. The rest of the blast will be directed straight up and out of the hole. Dig the grenade sumps as wide as the entrenching tool blade; at least as deep as an entrenching tool and as long as the position floor is wide (Figure 6-12).
7. Dig a storage compartment in the bottom of the back wall; the size of the compartment depends on the amount of equipment and ammunition to be stored (Figure 6-13).

 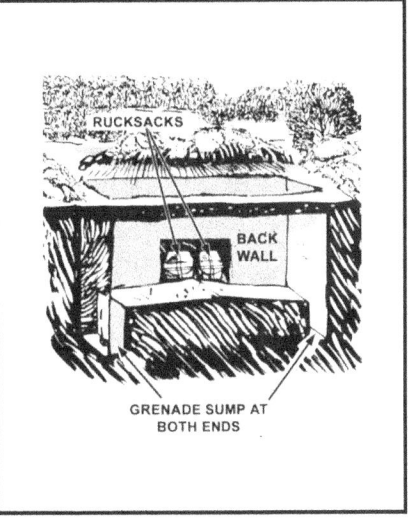

Figure 6-12. Grenade sumps.   Figure 6-13. Storage compartments.

8. Install revetments to prevent wall collapse/cave-in:
    a. Required in unstable soil conditions.
    b. Use plywood or sheeting material and pickets to revet walls.
    c. Tie back pickets and posts.
    d. Emplace OHC stringers:
    e. Use 2x4s, 4x4s, or pickets ("U" facing down).
    f. Make OHC stringers standard length, which is 8 feet (2.4 meters). This is long enough to allow sufficient length in case walls slope.
    g. Use "L" for stringer length and "H" for stringer spacing.
9. Remove the second layer of sandbags in the front and rear retaining walls to make room for the stringers. Place the same sandbags on top of the stringers once you have the stringers properly positioned.

Chapter 6

Stage 4

6-22. Install OHC and camouflage (Figure 6-14 and Figure 6-15):
1. Install overhead cover
2. Use plywood, sheeting mats as a dustproof layer (could be boxes, plastic panel, or interlocked U-shaped pickets). Standard dustproof layer is 4'x4' sheets of ¾-inch plywood centered over dug position.
3. Nail plywood dustproof layer to stringers.
4. Use at least 18 inches (46 centimeters) of sand-filled sandbags for overhead burst protection (four layers). At a minimum, these sandbags must cover an area that extends to the sandbags used for the front and rear retaining walls.
5. Use plastic or a poncho for waterproofing layer.
6. Fill center cavity with soil from dug hold and surrounding soil.
7. Use surrounding topsoil and camouflage screen systems.
8. Use soil from hole to fill sandbags, OHC cavity, and blend in with surroundings.

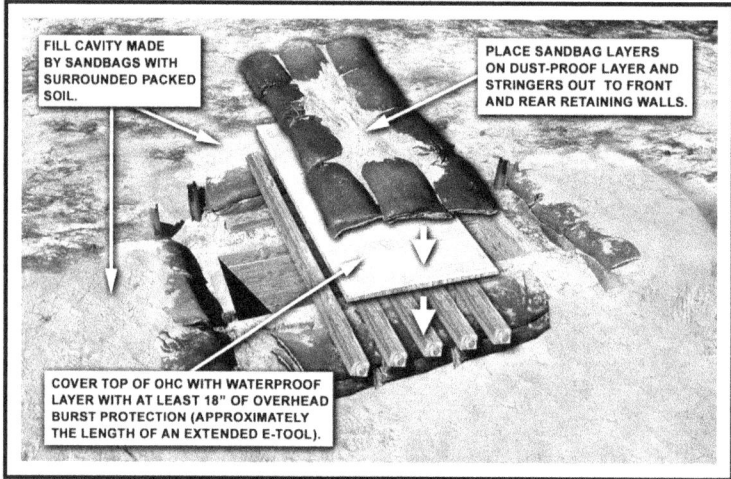

Figure 6-14. Installation of overhead cover.

Fighting Positions

Figure 6-15. Two-man fighting position with built-up OHC (Stage 4).

Built-Down Overhead Cover

6-23.   This should not exceed 12 inches (30 centimeters). This lowers the profile of the fighting position, which aids in avoiding detection. Unlike a built-up OHC, a built-down OHC has the following traits (Table 6-3, Figure 6-16, and Figure 6-17):

Chapter 6

Table 6-3. Specifications for built-down overhead cover.

Maximum 12 inches (30 centimeters) High
- You can build parapets up to 30 centimeters. Taper the overhead portions and parapets above the ground surface to conform to the natural lay of the ground.

Minimum Three M16s Long
- This gives you adequate fighting space between the end walls of the fighting position and the overhead cover. This takes 2.5 hours longer to dig in normal soil conditions.

Firing Platform for Elbows
- You must construct a firing platform in the natural terrain upon which to rest your elbows. The firing platform will allow the use of the natural ground surface as a grazing fire platform.

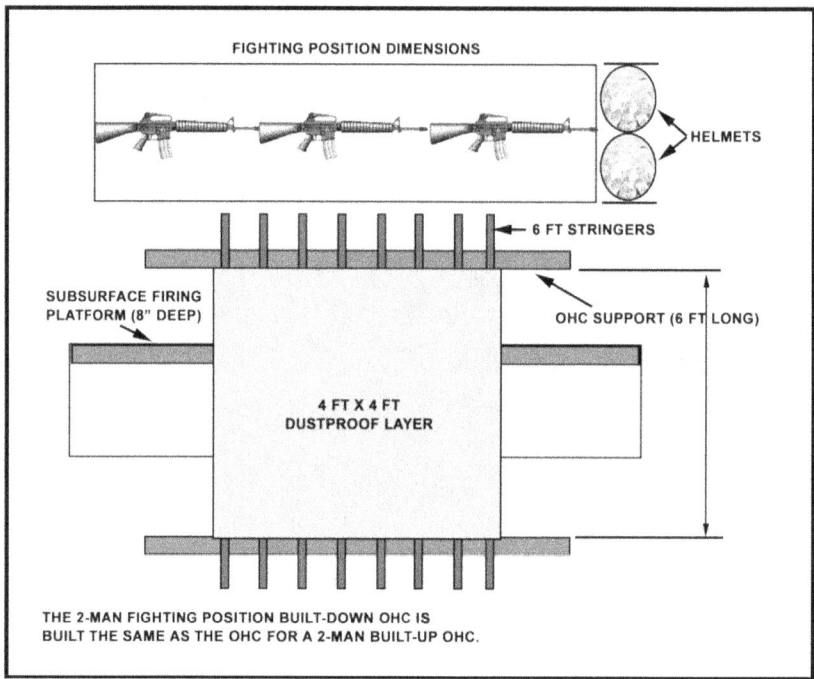

Figure 6-16. Two-man fighting position with built-down OHC (top view).

Fighting Positions

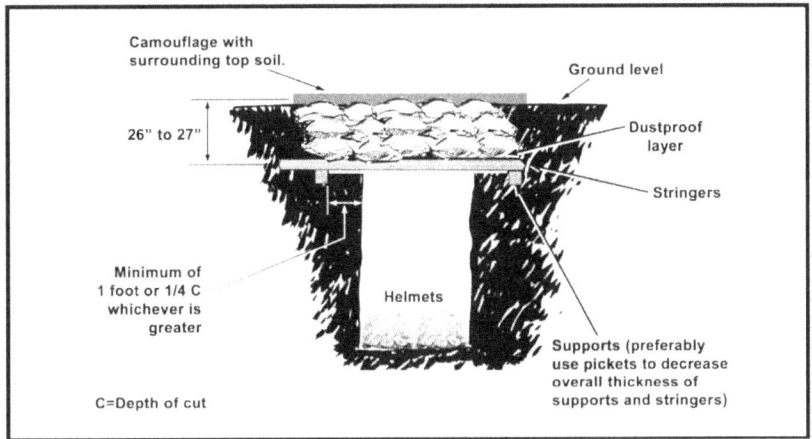

Figure 6-17. Two-man fighting position with built-down OHC (side view).

ONE-MAN FIGHTING POSITION

6-24. Sometimes you may have to build and occupy a one-man fighting position, for example, an ammunition bearer in a machine gun team. Except for its size, a one-man position is built the same way as a two-man fighting position. The hole of a one-man position is only large enough for you and your equipment. It does not have the security of a two-person position; therefore, it must allow a Soldier to shoot to the front or oblique from behind frontal cover.

MACHINE GUN FIGHTING POSITION

6-25. Construct fighting positions for machine guns so the gun fires to the front or oblique. However, the primary sector of fire is usually oblique so the gun can fire across your unit's front. Two Soldiers (gunner and assistant gunner) are required to Soldier the weapon system. Therefore, the hole is shaped so both the gunner and assistant gunner can get to the gun and fire it from either side of the frontal protection. The gun's height is reduced by digging the tripod platform down as much as possible. However, the platform is dug to keep the gun traversable across the entire sector of fire. The tripod is used on the side with the primary sector of fire, and the bipod legs are used on the side with the secondary sector. When changing from primary to secondary sectors, the machine gun is moved but the tripod stays in place. With a three-Soldier crew for a machine gun, the (ammunition bearer) digs a one-Soldier fighting position to the flank. From this position, the Soldier can see and shoot to the front and oblique. The ammunition bearer's position is connected to the gun position by a crawl trench so the bearer can transport ammunition or replace one of the gunners.

6-26. When a machine gun has only one sector of fire, dig only half of the position. With a three-man crew, the third Soldier (the ammunition bearer) digs a one-man fighting position. A one-man position is built the same as a two-man fighting position. The hole of a one-man position is only large enough for you and your equipment. Usually, his position is on the same side of the machine gun as its FPL or PDF. From that position, he can observe and fire into the machine gun's secondary sector and, at the same time, see the gunner and assistant gunner. The ammunition bearer's position is connected to the machine gun position by a crawl trench so that he can bring ammunition to the gun or replace the gunner or the assistant gunner.

Stage 1

6-27. Establish sectors (primary and secondary) of fire, and then outline position (Figure 6-18):

1. Check fields of fire from prone.
2. Assign sector of fire (primary and secondary) and final protective line (FPL) or principal direction of fire (PDF).
3. Emplace aiming stakes.
4. Decide whether to build OHC up or down, based on potential enemy observation of position.
5. Trace position outline to include location of two distinct firing platforms.
6. Mark position of the tripod legs where the gun can be laid on the FPL or PDF.
7. Clear primary and secondary fields of fire.

Note: The FPL is a line on which the gun fires grazing fire across the unit is front. Grazing fire is fired 1 meter above the ground. When an FPL is not assigned, a PDF is assigned. A PDF is a direction toward which the gun must be pointed when not firing at targets in other parts of its sector.

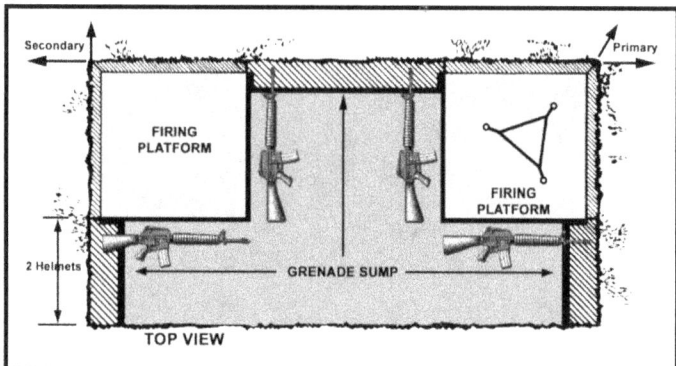

Figure 6-18. Position with firing platforms.

Stage 2

6-28. Dig firing platforms and emplace supports for OHC stringers, and then construct the parapet retaining walls:

1. Emplace OHC supports to front and rear of position.
2. Center OHC in position, and place supports as you did for Stage 2, two-man fighting position.
3. Construct the same as you did for Stage 2, two-man fighting position.
4. Dig firing platforms 6 to 8 inches (15 to 20 centimeters) deep and then position machine gun to cover primary sector of fire.
5. Use soil to fill sandbags for walls.

Fighting Positions

Stage 3

6-29. Dig position and build parapets, and then place stringers for the OHC (Figure 6-19):
1. Dig the position to a maximum armpit depth around the firing platform.
2. Use soil from hole to fill parapets in order of front, flanks, and rear.
3. Dig grenade sumps and slope floor toward them.
4. Install revetment if needed.
5. Follow same steps as for two-man fighting position.
6. Place stringers for OHC.
7. Follow same steps established for two-man fighting position.
8. Make stringers at least 8 feet (2.44 meters) long.

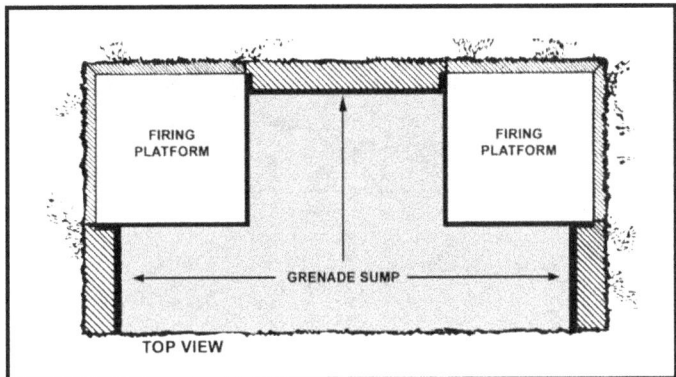

Figure 6-19. Grenade sump locations.

Stage 4

6-30. Install overhead cover (OHC) and camouflage (Figure 6-20):
1. For a machine gun position, build the OHC the same as you would for a two-man fighting position.
2. Use surrounding topsoil and camouflage screen systems.
3. Ensure no enemy observation within 115 feet (35 meters) of position.
4. Use soil from hole to fill sandbags and OHC cavity, or to spread around and blend position in with surrounding ground.

Chapter 6

Figure 6-20. Machine gun fighting position with OHC.

CLOSE COMBAT MISSILE FIGHTING POSITIONS

6-31. The following paragraphs discuss close combat missile fighting positions for the AT4 and Javelin:

AT4 POSITION

6-32. The AT4 is fired from the fighting positions previously described. However, backblast may cause friendly casualties of Soldiers in the position's backblast area. You should ensure that any walls, parapets, large trees, or other objects to the rear will not deflect the backblast. When the AT4 is fired from a two-Soldier position, you must ensure the backblast area is clear. The front edge of a fighting position is a good elbow rest to help you steady the weapon and gain accuracy. Stability is better if your body is leaning against the position's front or side wall.

Fighting Positions

STANDARD JAVELIN FIGHTING POSITION WITH OVERHEAD COVER

6-33. The standard Javelin fighting position has cover to protect you from direct and indirect fires (Figure 6-21). The position is prepared the same as the two-man fighting position with two additional steps. First, the back wall of the position is extended and sloped rearward, which serves as storage area. Secondly, the front and side parapets are extended twice the length as the dimensions of the two-man fighting position with the javelin's primary and secondary seated firing platforms added to both sides.

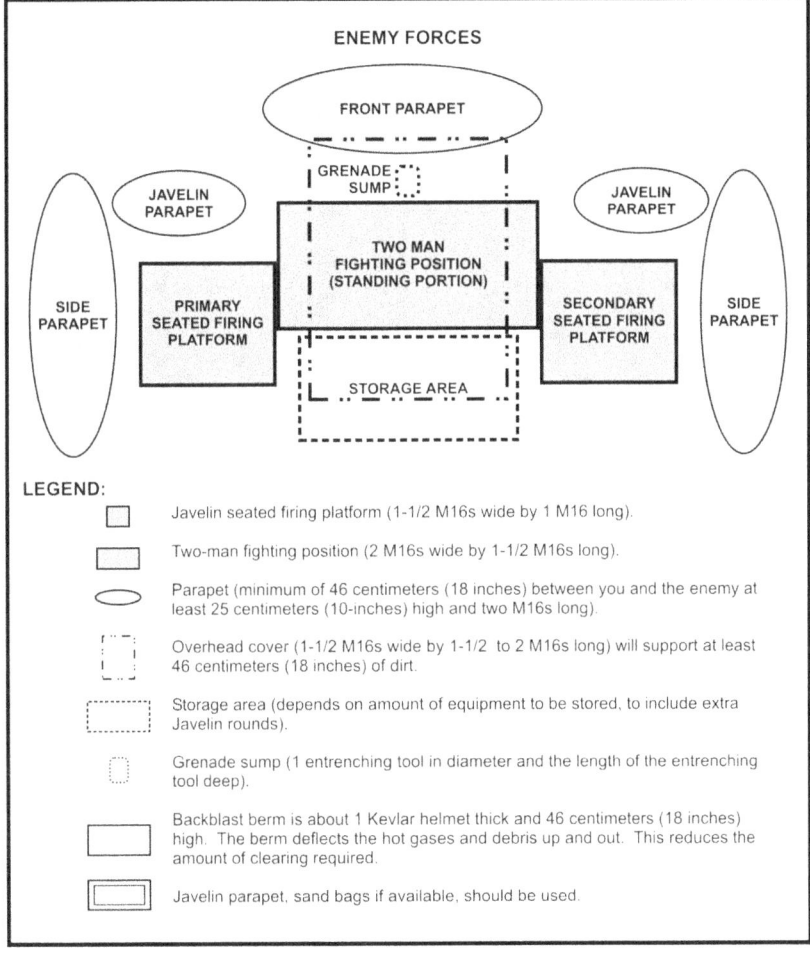

Figure 6-21. Standard Javelin fighting position.

13 August 2013 TC 3-21.75 6-21

Chapter 6

> *Note*: When a Javelin is fired, the muzzle end extends 6 inches (15 centimeters) beyond the front of the position, and the rear launcher extends out over the rear of the position. As the missile leaves the launcher, stabilizing fins unfold. You must keep the weapon at least 6 inches (15 centimeters) above the ground when firing to leave room for the fins. OHC that would allow firing from beneath it is usually built if the backblast area is clear.

RANGE CARDS

6-34. A range card, a rough plan of the terrain around a weapon position, is a sketch of the assigned sector that a direct fire weapon system is intended to cover. Range cards are prepared immediately upon arrival in a position, regardless of the length of stay, and updated as necessary. Two copies of the range card are prepared. One copy stays at your position and the other is sent to the platoon headquarters.

COMPONENTS

6-35. A range card is comprised of the following.

Sectors of Fire--A sector of fire is an area to be covered by fire that is assigned to an individual, a weapon, or a unit. You are normally assigned a primary and secondary sector of fire. Fire into your secondary sector of fire only if your primary sector has no targets, or if ordered to do so. Your gun's primary sector includes a FPL and a PDF.

Principal Direction of Fire --A PDF is a direction of fire assigned priority to cover an area that has good fields of fire or has a likely dismounted avenue of approach. The gun is positioned to fire directly down this approach rather than across the platoon's front. It also provides mutual support to an adjacent unit. Machine guns are sighted using the PDF if an FPL has not been assigned. If a PDF is assigned and other targets are not being engaged, machine guns remain on the PDF.

Final Protective Line--An FPL is a predetermined line along which grazing fire is placed to stop an enemy assault. Where terrain allows, your leader assigns an FPL to your weapon. An FPL becomes the machine gun's part of the unit is final protective fires. The FPL will be assigned to you only if your leader determines there is a good distance of grazing fire. If there is, the FPL will then dictate the location of the primary sector. The FPL will become the primary sector limit (right or left) closest to friendly troops. When not firing at other targets, you will lay your gun on the FPL or PDF.

Dead Space--Dead space is an area that direct fire weapons cannot hit. The area behind houses and hills, within orchards or defilades for example, is dead space. The extent of grazing fire and dead space may be determined in two ways. In the preferred method, the machine gun is adjusted for elevation and direction. Your assistant gunner walks along the FPL while you aim through the sights. In places where his waist (midsection) falls below your point of aim, dead space exists. Arm-and-hand signals must be used to control the Soldier who is walking and to obtain an accurate account of the dead space and its location. Another method is to observe the flight of tracer ammunition from a position behind and to the flank of the weapon.

AUTOMATIC WEAPON RANGE CARD

6-36. To prepare this range card--

1. Orient the card so both the primary and secondary sectors of fire (if assigned) can fit on it.
2. Draw a rough sketch of the terrain to the front of your position. Include any prominent natural and man-made features that could be likely targets.
3. Draw your position at the bottom of the sketch. Do not put in the weapon symbol at this time.
4. Fill in the marginal data to include--
5. Gun number (or squad).

6. Unit (only platoon and company) and date.
7. Magnetic north arrow.
8. Use the lensatic compass to determine magnetic north; and sketch in the magnetic north arrow on the card with its base starting at the top of the marginal data section.
9. Determine the location of your gun position in relation to a prominent terrain feature, such as a hilltop, road junction, or building. If no feature exists, place the eight-digit map coordinates of your position near the point where you determined your gun position to be. If there is a prominent terrain feature within 1,094 yards (1,000 meters) of the gun, use that feature. Do not sketch in the gun symbol at this time.
10. Using your compass, determine the azimuth in degrees from the terrain feature to the gun position. (Compute the back azimuth from the gun to the feature by adding or subtracting 180 degrees.)
11. Determine the distance between the gun and the feature by pacing or plotting the distance on a map.
12. Sketch in the terrain feature on the card in the lower left or right hand corner (whichever is closest to its actual direction on the ground) and identify it.
13. Connect the sketch of the position and the terrain feature with a barbed line from the feature to the gun.
14. Write in the distance in meters (above the barbed line).
15. Write in the azimuth in degrees from the feature to the gun (below the barbed line).

Final Protective Fires

6-37. To add an FPL to your range card (Figure 6-22):

1. Sketch in the limits of the primary sector of fire as assigned by your leader.
2. Sketch in the FPL on your sector limit as assigned.
3. Determine dead space on the FPL by having your AG walk the FPL. Watch him walk down the line and mark spaces that cannot be grazed.
4. Sketch dead space by showing a break in the symbol for an FPL, and write in the range to the beginning and end of the dead space.
5. Label all targets in your primary sector in order of priority. The FPL is number one.

Chapter 6

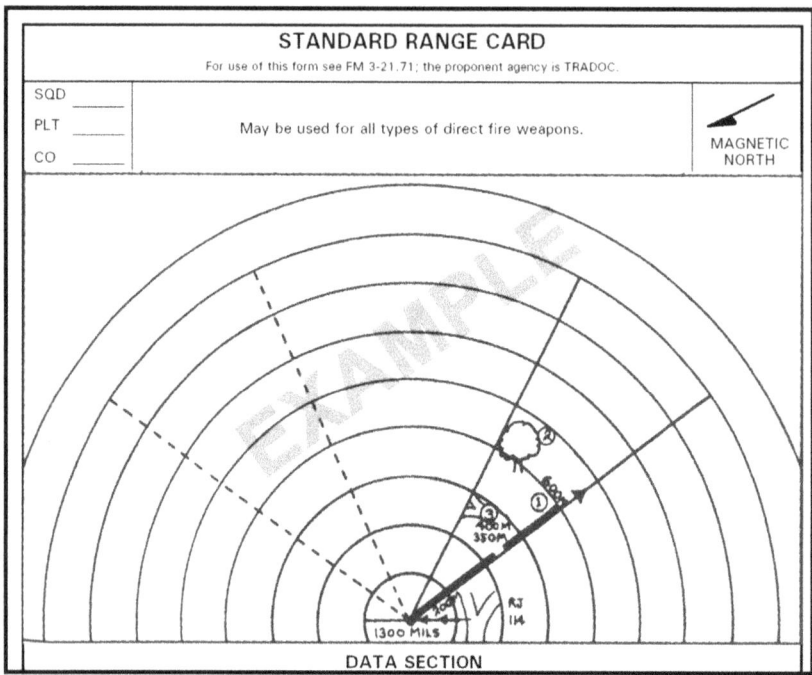

Figure 6-22. Primary sector with an FPL.

Primary Direction of Fire

6-38. To prepare your range card when assigned a PDF instead of an FPL (Figure 6-23):

1. Sketch in the limits of the primary sector of fire as assigned by your leader (sector should not exceed 875 mils, the maximum traverse of the tripod-mounted machine gun).
2. Sketch in the symbol for an automatic weapon oriented on the most dangerous target within your sector (as designated by your leader). The PDF will be target number one in your sector. All other targets will be numbered in priority.
3. Sketch in your secondary sector of fire (as assigned) and label targets within the secondary sector with the range in meters from your gun to each target. Use the bipod when it is necessary to fire into your secondary sector. The secondary sector is drawn using a broken line. Sketch in aiming stakes, if used.

Fighting Positions

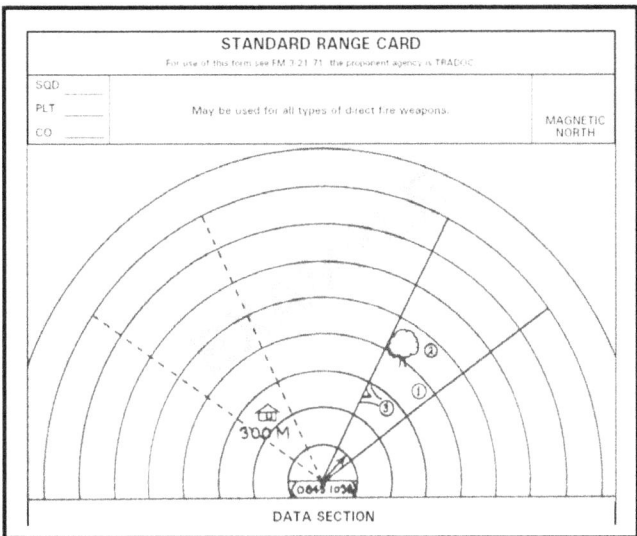

Figure 6-23. Complete sketch with PDF.

Data Section

6-39. The data section (Figure 6-24) of the range card lists the data necessary to engage targets identified in the sketch. The sketch does not have to be to scale, but the data must be accurate. The data section of the card can be placed on the reverse side or below the sketch if there is room. (Figure 6-25 shows an example completed data section.) Draw a data section block (if you do not have a printed card) with the following items:

Chapter 6

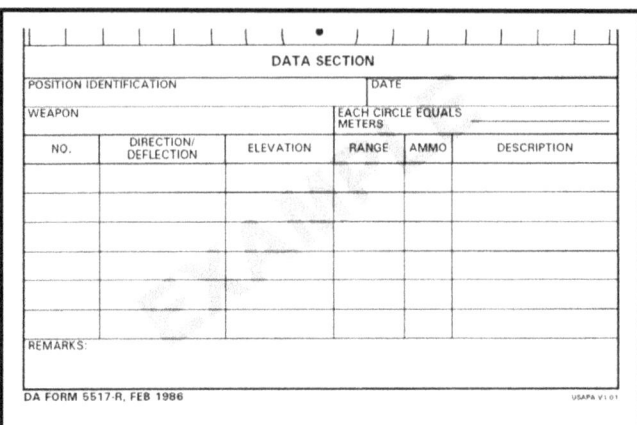

Figure 6-24. Data section.

Prepare

1. Center the traversing hand wheel.
2. Lay the gun for direction.
3. When assigned an FPL, lock the traversing slide on the extreme left or right of the bar, depending on which side of your primary sector the FPL is on.
4. Align the barrel on the FPL by moving the tripod legs. Do not enter a direction in the data section for the FPL.
5. When assigned a PDF, align your gun on the primary sector by traversing the slide to one side and then move the tripod to align the barrel on your sector limit. Align the PDF by traversing the slide until your gun is aimed at the center of the target.
6. Fix the tripod legs in place by digging in or sandbagging them. Once you emplace the tripod to fire into the primary sector, leave it there--do not move it.

Read Direction to Target

1. Lay your gun on the center of the target.
2. Read the direction directly off the traversing bar at the left edge of the traversing bar slide.
3. Enter the reading under the direction column of your range card data section.
4. Determine the left or right reading based on the direction of the barrel, just the opposite of the slide.
5. Lay your gun on the base of the target by rotating the elevating handwheel.
6. Read the number, including a plus or minus sign, except for "0" above the first visible line on the elevating scale. The sketch reads "–50."
7. Read the number on the elevating handwheel that is in line with the indicator. The sketch reads "3."
8. Enter this reading under the ELEVATION column of your range card data section. Separate the two numbers with a solidus, also known as a slash ("/"). Always enter the reading from the upper elevating bar first. The sketch reads "–50/3."
9. Enter the range to each target under the appropriate column in the data section.
10. Enter your ammunition type under the appropriate column in the data section.
11. Describe each target under the appropriate column in the data section.

Complete Remarks Section

1. Enter the width and depth of linear targets in mils. The "-4" means that if you depress the barrel 4 mils, the strike of the rounds will go down to ground level along the FPL.
2. When entering the width of the target, be sure to give the width in mils, and express it as two values. For example, the illustration shows that target number three has a width of 15 mils. The second value, L7, means that once the gun is laid on your target, traversing 7 mils to the LEFT will lay the gun on the left edge of the target.
3. Enter aiming stake if one is used for the target.
4. No data for the secondary sector will be determined since your gun will be fired in the bipod role.

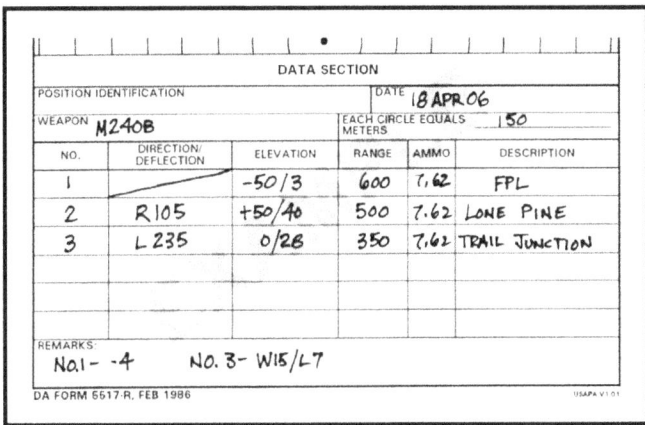

Figure 6-25. Example completed data section.

CLOSE COMBAT MISSILE RANGE CARD

6-40. The purpose of this card is to show a sketch of the terrain a weapon has been assigned to cover by fire. By using a range card, you can quickly and accurately determine the information needed to engage targets in your assigned sector (Figure 6-26 for a completed range card). Before you prepare a range card, your leader will show you where to position your weapon so you can best cover your assigned sector of fire. He will then, again, point out the terrain you are to cover. He will do this by assigning you a sector of fire or by assigning left or right limits indicated by either terrain features or azimuths. If necessary, he may also assign you more than one sector of fire and will designate the sectors as primary and secondary.

Chapter 6

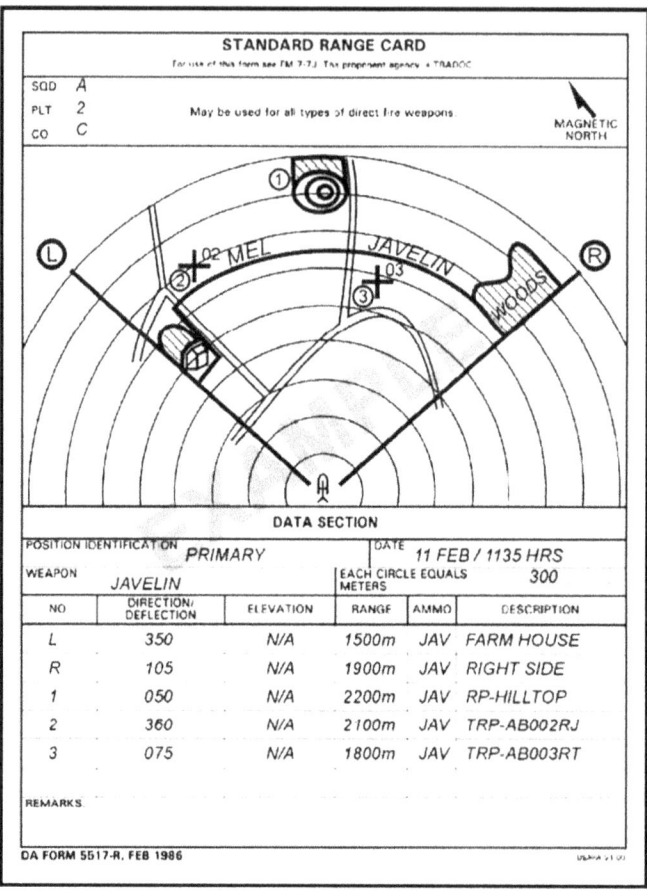

Figure 6-26. Example completed range card.

Target Reference Points

6-41. TRPs are natural or man-made features within your sector that you can use to quickly locate targets (Figure 6-27). TRPs are used mainly to control *direct fire* weapons. However, TRPs should appear on the company target list.

Maximum Engagement Line

6-42. The maximum engagement line (MEL) is a line beyond which you cannot engage a target. This line may be closer than the maximum engagement range of your weapon. Both the terrain and the maximum engagement range of your weapon will determine the path of the MEL.

Preparation

6-43. Draw the weapon symbol in the center of the small circle.

Sector Limits

6-44. Draw two lines from the position of the weapons system extending left and right to show the limits of the sector. The area between the left and right limits depicts your sector of fire or area of responsibility. Number the left limit as No. 1, number the right limit No. 2, and place a circle around each number. Record the azimuth and distance of each limit in the data section. Determine the value of each circle by finding a terrain feature farthest from the position and within the weapon system's capability. Determine the distance to the terrain feature. Round off the distance to the next even hundredth, if necessary. Determine the maximum number of circles that will divide evenly into the distance. The result is the value of each circle. Draw the terrain feature on the appropriate circle on the range card. Clearly mark the increment for each circle across the area where DATA SECTION is written. For example, suppose you use a hilltop at 2,565 yards (2,345 meters). Round the distance to 2,625 yards (2,400 meters) and divide by 8. The result is 300, so now each circle has a value of 300 meters.

Reference Points

Draw all reference points (RP) and TRPs in the sector. Mark each with a circled number beginning with 1. Draw hilltop as RP1, a road junction as RP2, and road junction RP3. Sometimes, a TRP and RP are the same point such as in the previous example. When this happens, mark the TRP with the first designated number in the upper right quadrant, and mark the RP in the lower left quadrant of the cross. This occurs when a TRP is used for target acquisition and range determination.

Road Junction--For a road junction, first determine the range to the junction, then draw the junction, and then draw the connecting roads from the road junction.

Dead Space--Show dead space as an irregular circle with diagonal lines inside. Any object that prohibits observation or coverage with direct fire will have the circle and diagonal lines extend out to the farthest MEL. If you can engage the area beyond the dead space, then close the circle.

Chapter 6

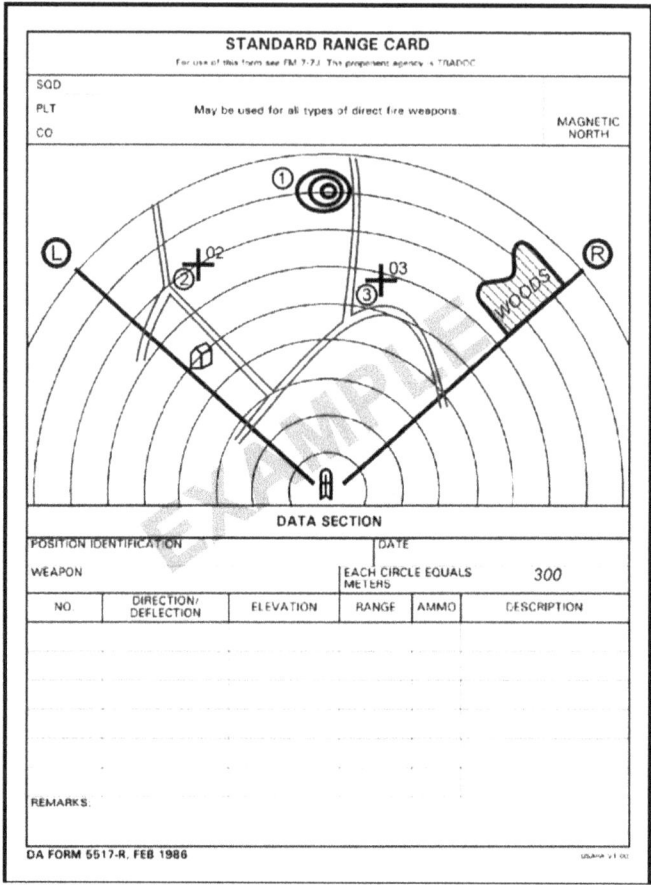

Figure 6-27. Reference points and target reference points.

Fighting Positions

Maximum Engagement Line

6-45. Draw the MEL at the maximum effective engagement range for the weapon, but draw it around (inside) the near edge of any dead spaces (Figure 6-28). *Do not* draw the MEL through dead spaces.

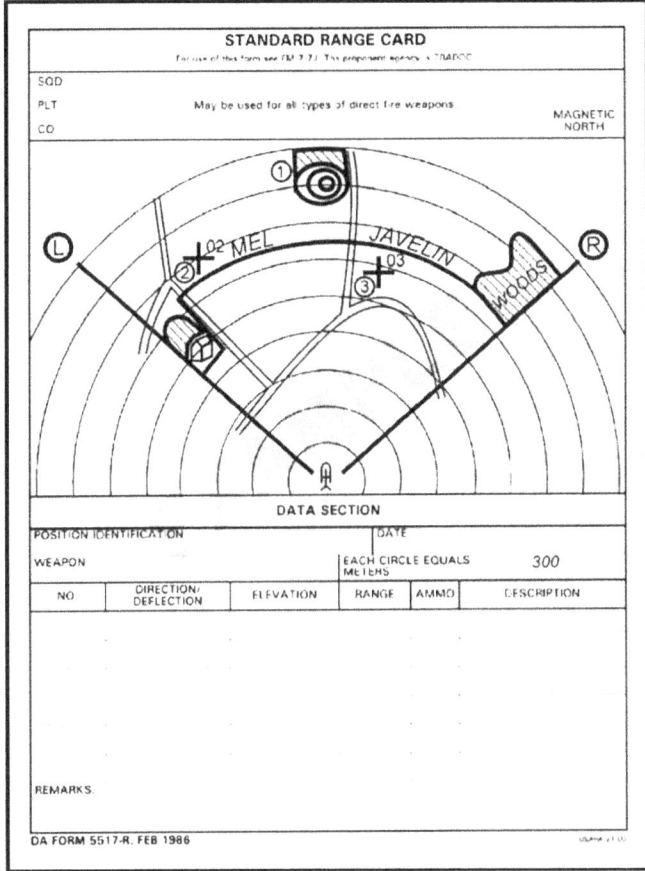

Figure 6-28. Maximum engagement lines.

Chapter 6

Weapon Reference Point

6-46. Show the WRP as a line with a series of arrows, extending from a known terrain feature, and pointing in the direction of the weapon system symbol (Figure 6-29). Number this feature last. The WRP location is given a six-digit grid. When there is no terrain feature to be designated as the WRP, show the weapon's location as an eight-digit grid coordinate in the Remarks block of the range card. Complete the data section as follows:

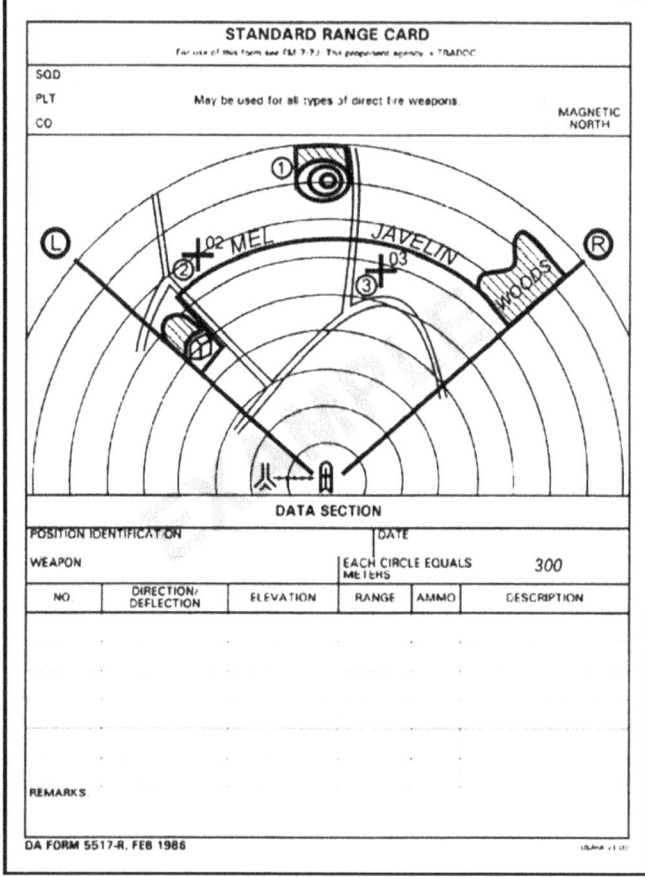

Figure 6-29. Weapon reference point.

Fighting Positions

Position Identification--List primary, alternate, or supplementary positions. Alternate and supplemental positions must be clearly identified.

Date--Show date and time the range card was completed. Range cards, like fighting positions, are constantly updated. The date and time are vital in determining current data.

Weapon--The weapon block indicates weapon type.

Each Circle Equals ____ Meters--Write in the distance, in meters, between circles.

NO (Number)--Start with L and R limits, then list TRPs and RPs in numerical order.

Direction/Deflection--The direction is listed in degrees. The deflection is listed in mils.

Elevation--The elevation is listed in mils.

Range--This is the distance, in meters, from weapon system position to L and R limits and TRPs and RPs.

Ammunition--List types of ammunition used.

Description--List the name of the object (for example, farmhouse, wood line, or hilltop).

Remarks--Enter the WRP data. As a minimum, WRP data describes the WRP and gives its six-digit or eight digit grid coordinate, magnetic azimuth, and distance to the position. Complete the marginal information at the top of the card.

Unit Description--Enter unit description such as squad, platoon, or company. Never indicate a unit higher than company.

Magnetic North--Orient the range card with the terrain, and draw the direction of the magnetic North arrow.

This page intentionally left blank.

Chapter 7

Movement

Normally, you will spend more time moving than fighting. The fundamentals of movement discussed in this chapter provide techniques that all Soldiers must learn. Even seasoned troops should practice these techniques regularly, until they become second nature.

INDIVIDUAL MOVEMENT TECHNIQUES

7-1. Your leaders base their selection of a particular movement technique by traveling, traveling overwatch, or bounding overwatch on the likelihood of enemy contact and the requirement for speed. However, your unit is ability to move depends on your movement skills and those of your fellow Soldiers. Use the following techniques to avoid being seen or heard:

- Stop, look, listen, and smell (SLLS) before moving. Look for your next position before leaving a position.
- Look for covered and concealed routes on which to move.
- Change direction slightly from time-to-time when moving through tall grass.
- Stop, look, and listen when birds or animals are alarmed (the enemy may be nearby).
- Smell for odors such as petroleum, smoke, and food; they are additional signs of the enemy's presence.
- Cross roads and trails at places that have the most cover and concealment (large culverts, low spots, curves, or bridges).
- Avoid steep slopes and places with loose dirt or stones.
- Avoid cleared, open areas and tops of hills and ridges. Walking at the top of a hill or ridge will skyline you against the sun or moon, enabling the enemy to see you.

INDIVIDUAL MOVEMENT TECHNIQUES

7-2. In addition to walking, you may move in one of three other methods known as individual movement techniques (IMT) — low crawl, high crawl, or rush.

Low Crawl

7-3. The low crawl gives you the lowest silhouette. Use it to cross places where the cover and/or concealment are very low and enemy fire or observation prevents you from getting up. Keep your body flat against the ground. With your firing hand, grasp your weapon sling at the upper sling swivel. Let the front hand guard rest on your forearm (keeping the muzzle off the ground), and let the weapon butt drag on the ground. To move, push your arms forward and pull your firing side leg forward. Then pull with your arms and push with your leg. Continue this throughout the move (Figure 7-1).

Chapter 7

High Crawl

7-4. The high crawl lets you move faster than the low crawl and still gives you a low silhouette. Use this crawl when there is good cover and concealment but enemy fire prevents you from getting up. Keep your body off the ground and resting on your forearms and lower legs. Cradle your weapon in your arms and keep its muzzle off the ground. Keep your knees well behind your buttocks so your body will stay low. To move, alternately advance your right elbow and left knee, then your left elbow and right knee (Figure 7-1).

Figure 7-1. Low and high crawl.

7-5. When you are ready to stop moving:
- Plant both of your feet.
- Drop to your knees (at the same time slide a hand to the butt of your rifle).
- Fall forward, breaking the fall with the butt of the rifle.
- Go to a prone firing position.

7-6. If you have been firing from one position for some time, the enemy may have spotted you and may be waiting for you to come up from behind cover. So, before rushing forward, roll or crawl a short distance from your position. By coming up from another spot, you may fool an enemy who is aiming at one spot and waiting for you to rise. When the route to your next position is through an open area, use the 3 to 5 second rush. When necessary, hit the ground, roll right or left, and then rush again.

Rush

7-7. The rush is the fastest way to move from one position to another (Figure 7-2). Each rush should last from 3 to 5 seconds. Rushes are kept short to prevent enemy machine gunners or riflemen from tracking you. However, do not stop and hit the ground in the open just, because 5 seconds have passed. Always try to hit the ground behind some cover. Before moving, pick out your next covered and concealed position and the best route to it. Make your move from the prone position as follows:
- Slowly raise your head and pick your next position and the route to it.
- Slowly lower your head.

Movement

- Draw your arms into your body (keeping your elbows in).
- Pull your right leg forward.
- Raise your body by straightening your arms.
- Get up quickly.
- Rush to the next position.

Figure 7-2. Rush.

Movement With Stealth

7-8. Moving with stealth means moving quietly, slowly, and carefully. This requires great patience. To move with stealth, use the following techniques:

- Ensure your footing is sure and solid by keeping your body's weight on the foot on the ground while stepping.
- Raise the moving leg high to clear brush or grass.

Chapter 7

- Gently let the moving foot down toe first, with your body's weight on the rear leg.
- Lower the heel of the moving foot after the toe is in a solid place.
- Shift your body's weight and balance the forward foot before moving the rear foot.
- Take short steps to help maintain balance.

7-9. At night, and when moving through dense vegetation, avoid making noise. Hold your weapon with one hand, and keep the other hand forward, feeling for obstructions. When going into a prone position, use the following techniques:

- Hold your rifle with one hand and crouch slowly.
- Feel for the ground with your free hand to make sure it is clear of mines, tripwires, and other hazards.
- Lower your knees, one at a time, until your body's weight is on both knees and your free hand.
- Shift your weight to your free hand and opposite knee.
- Raise your free leg up and back, and lower it gently to that side.
- Move the other leg into position the same way.
- Roll quietly into a prone position.

7-10. Use the following techniques when crawling:

- Crawl on your hands and knees.
- Hold your rifle in your firing hand.
- Use your nonfiring hand to feel for and make clear spots for your hands and knees.
- Move your hands and knees to those spots, and put them down softly.

MOVEMENT WITHIN A TEAM

7-11. Movement formations are used for control, security, and flexibility. These formations are the actual arrangements for you and your fellow Soldiers in relation to each other.

Control

7-12. Every squad and Soldier has a standard position. You must be able to see your fire team leader. Fire team leaders must be able to see their squad leaders. Leaders control their units using arm-and-hand signals.

Security

7-13. Formations also provide 360-degree security and allow the weight of their firepower to the flanks or front in anticipation of enemy contact.

Flexibility

7-14. Formations do not demand parade ground precision. Your leaders must retain the flexibility needed to vary their formations to the situation. The use of formations allows you to execute battle drills more quickly and gives the assurance that your leaders and buddy team members are in their expected positions and performing the right tasks. You will usually move as a member of a squad/team. Small teams, such as Infantry fire teams, normally move in a wedge formation. Each Soldier in the team has a set position in the wedge, determined by the type of weapon he carries. That position, however, may be changed by the team leader to meet the situation. The normal distance between Soldiers is 10 meters. When enemy contact is possible, the distance between teams should be about 50 meters. In very open terrain such as the desert, the interval may increase. The distance between individuals and teams is determined by how much command and control the squad leader can still exercise over his teams and the team members (Figure 7-3).

Movement

Figure 7-3. Fire team wedge.

7-15. You may have to make a temporary change in the wedge formation when moving through close terrain. The Soldiers in the sides of the wedge close into a single file when moving in thick brush or through a narrow pass. After passing through such an area, they should spread out, forming the wedge again. You should not wait for orders to change the formation or the interval. You should change automatically and stay in visual contact with the other team members and the team leader. The team leader leads by setting the example. His standing order is, *FOLLOW ME AND DO AS I DO*. When he moves to the left, you should move to the left. When he gets down, you should get down. When visibility is limited, control during movement may become difficult. To aid control, for example, the helmet camouflage band has two, 1-inch horizontal strips of luminous tape sewn on it. Unit SOPs normally address the configuration of the luminous strips.

IMMEDIATE ACTIONS WHILE MOVING

7-16. This section furnishes guidance for the immediate actions you should take when reacting to enemy indirect fire and flares. These Warrior Drills are actions every Soldier and small unit should train for proficiency.

REACTING TO INDIRECT FIRE

7-17. If you come under indirect fire while moving, immediately seek cover and follow the commands and actions of your leader. He will tell you to run out of the impact area in a certain direction or will tell you to follow him (Figure 7-4). If you cannot see your leader, but can see other team members, follow them. If alone, or if you cannot see your leader or the other team members, run out of the area in a direction away from the incoming fire.

Figure 7-4. Following of team leader from impact area.

7-18. It is hard to move quickly on rough terrain, but the terrain may provide good cover. In such terrain, it may be best to take cover and wait for the fires to cease. After they stop, move out of the area quickly.

REACTING TO GROUND FLARES

7-19. The enemy puts out ground flares as warning devices. He sets them off himself or attaches tripwires to them for you to trip on and set off. He usually puts the flares in places he can watch (Figure 7-5).

Movement

Figure 7-5. Reaction to ground flares.

7-20. If you are caught in the light of a ground flare, flip up your NVD and move quickly out of the lighted area. The enemy will know where the ground flare is and will be ready to fire into that area. Move well away from the lighted area. While moving out of the area, look for other team members. Try to follow or join them to keep the team together.

Chapter 7

REACTING TO AERIAL FLARES

7-21. The enemy uses aerial flares to light up vital areas. They can be set off like ground flares; fired from hand projectors, grenade launchers, mortars, and artillery; or dropped from aircraft. If you hear the firing of an aerial flare while you are moving, flip up your NVD and hit the ground (behind cover if possible) while the flare is rising and before it bursts and illuminates. If moving where it is easy to blend with the background, such as in a forest, and you are caught in the light of an aerial flare, freeze in place until the flare burns out.

7-22. If you are caught in the light of an aerial flare while moving in an open area, immediately crouch low or lie down. If you are crossing an obstacle, such as a barbed-wire fence or a wall, and are caught in the light of an aerial flare, crouch low and stay down until the flare burns out. The sudden light of a bursting flare may temporarily wash out your NVD, blinding both you and the enemy. When the enemy uses a flare to spot you, he spoils his own night vision. To protect your night vision, flip up your NVD and close one eye while the flare is burning. When the flare burns out, the eye that was closed will still have its night vision and you can place your NVD back into operation (Figure 7-6).

Figure 7-6. Reaction to aerial flares.

FIRE AND MOVEMENT

7-23. When a unit makes contact with the enemy, it normally starts firing at and moving toward the enemy. Sometimes the unit may move away from the enemy. This technique is called fire and movement—

Movement

one element maneuvers (or moves) while another provides a base of fire. It is conducted to both close with and destroy the enemy, or to move away from the enemy to break contact with him.

7-24. The firing and moving takes place at the same time. There is a fire element and a movement element. These elements may be buddy teams, fire teams, or squads. Regardless of the size of the elements, the action is still fire and movement.

- The fire element covers firing at and suppressing the enemy. This helps keep the enemy from firing back at the movement element.
- The movement element moves either to close with the enemy or to reach a better position from which to fire at him. The movement element should not move until the fire element is firing.

7-25. Depending on the distance to the enemy position and on the available cover, the fire element and the movement element switch roles as needed to keep moving. Before the movement element moves beyond the supporting range of the fire element (the distance in which the weapons of the fire element can fire and support the movement element), it should take a position from which it can fire at the enemy. The movement element then becomes the next fire element and the fire element becomes the next movement element. If your team makes contact, your team leader should tell you to fire or to move. He should also tell you where to fire from, what to fire at, or where to move. When moving, use the low crawl, high crawl, or rush IMTs.

MOVEMENT ON VEHICLES

7-26. Soldiers can ride on the outside of armored vehicles; however, this is not done routinely. Therefore, as long as tanks and Infantry are moving in the same direction and contact is not likely, Soldiers may ride on tanks.

GUIDELINES FOR RIDING ON ALL ARMORED VEHICLES

7-27. The following must be considered before Soldiers mount or ride on an armored vehicle.

- When mounting an armored vehicle, Soldiers must always approach the vehicle from the front to get permission from the vehicle commander to mount. They then mount the side of the vehicle away from the coaxial machine gun and in view of the driver. Maintain three points of contact and only use fixed objects as foot and handholds. Do not use gun or optic system.
- If the vehicle has a stabilization system, the squad leader obtains verification from the vehicle commander that it is OFF before the vehicle starts to move.
- The Infantry must dismount as soon as possible when tanks come under fire or when targets appear that require the tank gunner to traverse the turret quickly to fire.
- All Soldiers must be alert for obstacles that can cause the tank to turn suddenly and for trees that can knock riders off the tank.

GUIDELINES FOR RIDING ON SPECIFIC ARMORED VEHICLES

7-28. The following information applies to specific vehicles.

- The M1 tank is not designed to carry riders easily. Riders must NOT move to the rear deck. Engine operating temperatures make this area unsafe for riders (Figure 7-7).
- One Infantry squad can ride on the turret. The Soldiers must mount in such a way that their legs cannot become entangled between the turret and the hull by an unexpected turret movement. Rope and equipment straps may be used as a field-expedient Infantry rail to provide secure handholds. Soldiers may use a snap link to assist in securing themselves to the turret.

Chapter 7

- Everyone must be to the rear of the smoke grenade launchers. This automatically keeps everyone clear of the coaxial machine gun and laser range finder.
- The Infantry must always be prepared for sudden turret movement.
- Leaders should caution Soldiers about sitting on the turret blowout panels, because 250 pounds of pressure will prevent the panels from working properly. If there is an explosion in the ammunition rack, these panels blow outward to lessen the blast effect in the crew compartment.
- If enemy contact is made, the tank should stop in a covered and concealed position, and allow the Infantry time to dismount and move away from the tank. This action needs to be rehearsed before movement.
- The Infantry should not ride with anything more than their battle gear. Excess gear should be transported elsewhere.

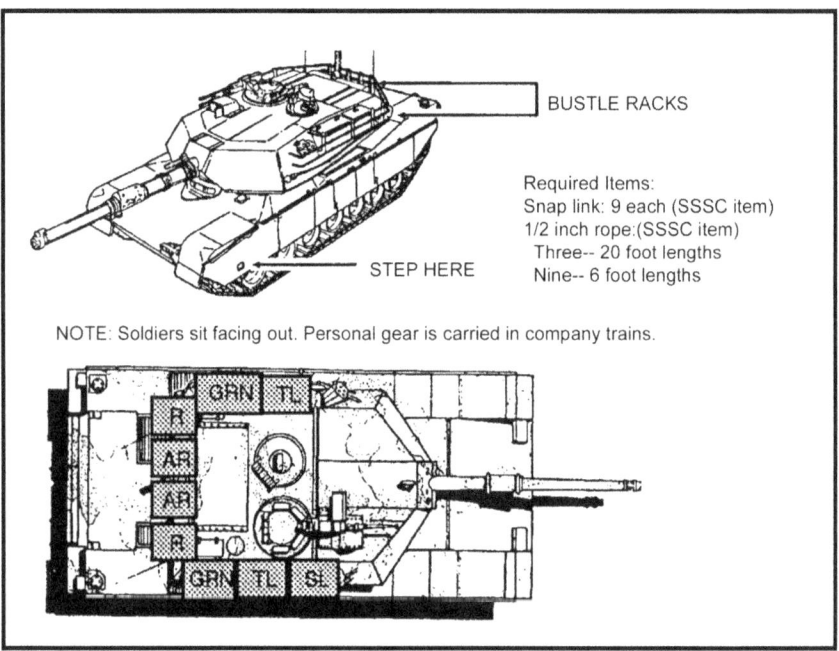

Figure 7-7. Mounting and riding arrangements.

Chapter 8

Urban Areas

The rapid growth of the number and size of urban centers, especially in regions of political instability, increases the likelihood that Soldiers will be called upon to conduct operations in urban areas. Keep in mind that the urban battlefield environment is rapidly exhausting, both physically and mentally, and may look even more chaotic than it is. Successful combat operations in urban areas require skills that are unique to this type of fighting. You must be skilled in moving, entering buildings, clearing rooms, and selecting and using fighting positions to be effective while operating in this type of environment

Section I. MOVEMENT TECHNIQUES

Movement in urban areas is the first skill you must master. Movement techniques must be practiced until they become second nature. To reduce exposure to enemy fire, you should avoid open areas, avoid silhouetting yourself, and select your next covered position before movement. The following paragraphs discuss how to move in urban areas:

AVOIDING OPEN AREAS

8-1.   Open areas, such as streets, alleys, and parks, should be avoided. They are natural kill zones for enemy, crew-served weapons, or snipers. They can be crossed safely if the individual applies certain fundamentals, including using smoke from hand grenades or smoke pots to conceal movement. When employing smoke as an obscurant, keep in mind that thermal sighting systems can see through smoke. Also, when smoke has been thrown in an open area, the enemy may choose to engage with suppressive fires into the smoke cloud.

- Before moving to another position, you should make a visual reconnaissance, select the position offering the best cover and concealment, and determine the route to get to that position.
- You need to develop a plan for movement. You should always select the shortest distance to run between buildings and move along covered and concealed routes to your next position, reducing the time exposed to enemy fire.

MOVING PARALLEL TO BUILDINGS

8-2.   You may not always be able to use the inside of buildings as routes of advance and must move on the outside of the buildings. Smoke, suppressive fires, and cover and concealment should be used as much as possible to hide movement. You should move parallel to the side of the building, maintaining at least 12 inches of separation between yourself and the wall to avoid rabbit rounds (ricochets and rubbing or bumping the wall). Stay in the shadows, present a low silhouette, and move rapidly to your next position. If an enemy gunner inside the building fires, he exposes himself to fire from other squad members providing overwatch.

Chapter 8

MOVING PAST WINDOWS

8-3. Windows present another hazard to the Soldier. The most common mistakes are exposing the head in a first-floor window and not being aware of basement windows. When using the correct technique for passing a first-floor window, you must stay below the window level and near the side of the building (Figure 8-1). Ensure you do not silhouette yourself in the window. An enemy gunner inside the building would have to expose himself to covering fires if he tries to engage you.

Figure 8-1. Soldier moving past windows.

8-4. The same techniques used in passing first-floor windows are used when passing basement windows. You should not walk or run pass a basement window, as this will present a good target for an enemy gunner inside the building. Ensure you stay close to the wall of the building and step or jump pass the window without exposing your legs (Figure 8-2).

Urban Areas

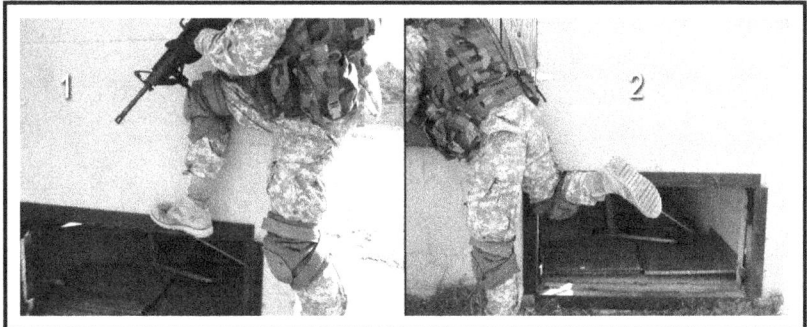

Figure 8-2. Soldier passing basement windows.

CROSSING A WALL

8-5.   You must learn the correct method of crossing a wall (Figure 8-3). After you have reconnoitered the other side, quickly roll over the wall and keep a low silhouette. Your speed of movement and low silhouette denies the enemy a good target.

Figure 8-3. Soldier crossing a wall.

MOVING AROUND CORNERS

8-6. The area around a corner must be observed before the Soldier moves. The most common mistake you can make at a corner is allowing your weapon to extend beyond the corner, exposing your position; this mistake is known as *flagging* your weapon. You should show your head below the height an enemy would expect to see it. You must lie flat on the ground and not extend your weapon beyond the corner of the building. Only expose your head (at ground level) enough to permit observation (Figure 8-4). You can also use a mirror, if available, to look around the corner. Another corner-clearing technique that is used when speed is required is the *pie-ing* method. This procedure is done by aiming the weapon beyond the corner into the direction of travel (without flagging) and side-stepping around the corner in a circular fashion with the muzzle as the pivot point (Figure 8-5).

Figure 8-4. Correct technique for looking around a corner.

Figure 8-5. *Pie-ing* a corner.

MOVING WITHIN A BUILDING

8-7. Once you have entered a building (see Section II), follow these procedures to move around in it:

DOORS AND WINDOWS

8-8. Avoid silhouetting yourself in doors and windows (Figure 8-6).

HALLWAYS

8-9. When moving in hallways, never move alone—always move with at least one other Soldier for security.

WALLS

8-10. You should try to stay 12 to 18 inches away from walls when moving; rubbing against walls may alert an enemy on the other side, or, if engaged by an enemy, ricochet rounds tend to travel parallel to a wall.

Figure 8-6. Movement within a building.

Chapter 8

Section II. OTHER PROCEDURES

This section discusses how to enter a building, clear a room, and use fighting positions.

ENTERING A BUILDING

8-11. When entering buildings, exposure time must be minimized. Before moving toward the building, you must select the entry point. When moving to the entry point use smoke to conceal your advance. You must avoid using windows and doors except as a last resort. Consider the use of demolitions, shoulder-launched munitions (SLMs), close combat missiles (CCMs), tank rounds, and other means to make new entrances. If the situation permits, you should precede your entry with a grenade, enter immediately after the grenade explodes, and be covered by one of your buddies. Entry should be made at the highest level possible.

ENTER UPPER LEVEL

8-12. Entering a building from any level other than the ground floor is difficult. However, clearing a building from the top down is best, because assaulting and defending are easier from upper floors. Gravity and the building's floor plan help when Soldiers throw hand grenades and move between floors. An enemy forced to the top of a building may be cornered and fight desperately, or escape over the roof. An enemy who is forced down to ground level may withdraw from the building, exposing himself to friendly fires from the outside. Soldiers can use several means, including ladders, drainpipes, vines, helicopters, or the roofs and windows of adjacent buildings, to reach the top floor or roof of a building. One Soldier can climb onto the shoulders of another and reach high enough to pull himself up. Ladders are the fastest way to reach upper levels. If portable ladders are unavailable, Soldiers can construct them from materials available through supply channels. They can also build ladders using resources available in the urban area. For example, they can use the lumber from inside the walls of buildings. Although ladders do not permit access to the top of some buildings, they do offer security and safety through speed. Soldiers can use ladders to conduct an exterior assault of an upper level, provided exposure to enemy fire can be minimized.

SCALE WALLS

8-13. When you must scale a wall during exposure to enemy fire, use all available concealment. Use smoke and other diversions to improve your chance of success. When using smoke for concealment, plan for wind direction. Use suppressive fire, shouting, and distractions from other positions to divert the enemy's attention. You are vulnerable to enemy fire when scaling an outside wall. Ideally, move from building to building and climb buildings only under cover of friendly fire. Properly positioned friendly weapons can suppress and eliminate enemy fire. If you must scale a wall with a rope, avoid silhouetting yourself in windows of uncleared rooms, and avoid exposing yourself to enemy fires from lower windows. Climb with your weapon slung over your firing shoulder so you can bring it quickly to a firing position. If the rules of engagement (ROE, which are the rules governing the use of force) permit, engage the objective window and any lower level windows in your path with grenades (hand or launcher) before you ascend. Enter the objective window with a low silhouette. You can enter head first, but the best way is to hook a leg over the window sill and enter sideways, straddling the ledge.

Urban Areas

ENTER AT LOWER LEVELS

8-14. Buildings are best cleared from the top down. However, you might not be able to enter a building from the top. Entry at the bottom or lower level is common, and might be the only way. When entering at lower levels, avoid entering through windows and doors, since either is easily booby trapped, and both are usually covered by enemy fire (Figure 8-7 Figure 8-8, this page; and Figure 8-9 and Figure 8-10 on page 8-8). Use these techniques when you can enter the building without receiving effective enemy fire. When entering at lower levels, use demolitions, artillery, tank fire, SLMs, CCMs, ramming of an armored vehicle into a wall, or similar means to create a new entrance and avoid booby traps. This is the best technique, ROE permitting. Once you use these means, enter quickly to take advantage of the effects of the blast and concussion. Door breaching is the best way to enter at the lower level. Before entering, you may throw a hand grenade into the new entrance to reinforce the effects of the original blast.

Note: Armored vehicles can be positioned next to a building, so Soldiers can use them as a platform for entering a room or gaining access to a roof.

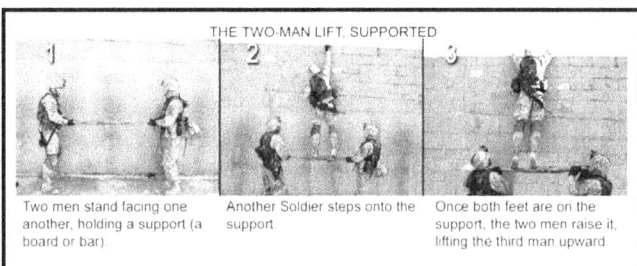

Figure 8-7. Lower-level entry technique with support bar.

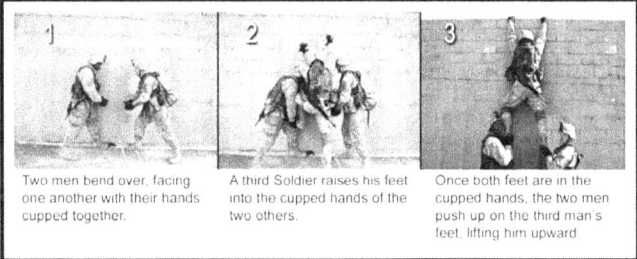

Figure 8-8. Lower-level entry technique without support bar.

Chapter 8

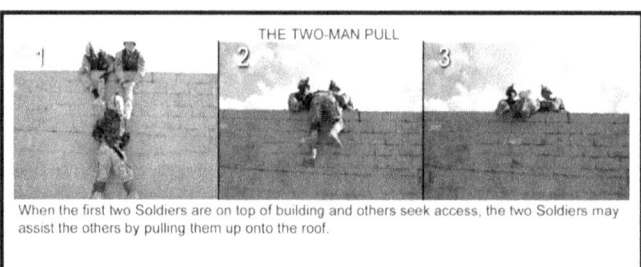

Figure 8-9. Lower-level entry two-man pull technique.

Figure 8-10. Lower-level entry one-man lift technique.

8-15. Blow or cut breach holes through walls to allow you to enter a building. Such entrances are safer than doors, because doors are easily booby trapped, and should be avoided, unless you conduct an explosive breach on the door.

- Throw a grenade through the breach before entering. Use available cover, such as the lower corner of the building, for protection from fragments.
- Use stun and concussion grenades when engaging through thin walls.

8-16. When a door is your only way into a building, beware of booby traps and fire from enemy soldiers inside the room. You can breach (force open) a locked door using one of four breaching methods:

- Mechanical.
- Ballistic.
- Explosive.
- Thermal.

8-17. If none of these methods is available, you may kick the door open. This is worst method, since it is difficult and tiring. Also, it rarely works the first time, giving any enemy inside ample time to shoot you through the door.

- When opening an unlocked door by hand, make sure you and the rest of the assault team avoid exposing themselves to enemy fire through the door. To reduce exposure, stay close to one side of the doorway.
- ROE permitting, once you get the door open, toss in a hand grenade. Once it explodes, enter and clear the room.

EMPLOY HAND GRENADES

8-18. Combat in urban areas often requires extensive use of hand grenades. Unless the ROE or orders prevent it, use grenades before assaulting defended areas, moving through breaches, or entering unsecured areas. Effective grenade use in urban areas may require throwing overhand or underhand, with either the left or right hand.

Note: To achieve aboveground detonation or near-impact detonation, remove the grenade's safety pin, release the safety lever, count "*One thousand one, one thousand two,*" and throw the grenade. This is called *cooking-off*. Cooking off takes about 2 seconds of the grenade's 4- to 5-second delay, and it allows the grenade to detonate above ground or shortly after impact with the target.

Types

8-19. Three types of hand grenades can be used when assaulting an urban objective: stun, concussion, and fragmentation. The type of construction materials used in the objective building influence the type of grenades that can be used.

- ***M84 Stun Hand Grenade*--**This grenade is a flash-bang distraction device that produces a brilliant flash and a loud bang to briefly surprise and distract an enemy force. The M84 is often used under precision conditions and when the ROE demand use of a nonlethal grenade. The use of stun hand grenades under high intensity conditions is usually limited to situations where fragmentation and concussion grenades pose a risk to friendly troops or the structural integrity of the building.
- ***Concussion Grenade*--**The concussion grenade causes injury or death to persons in a room by blast overpressure and propelling debris within the room. While the concussion grenade does not discard a dangerous fragmentation from its body, the force of the explosion can create debris fallout that may penetrate thin walls.
- ***Fragmentation Grenade*--**The fragmentation grenade produces substantial overpressure when used inside buildings, and coupled with the shrapnel effects, can be extremely dangerous to friendly Soldiers. If the walls of a building are made of thin material, such as sheetrock or thin plywood, you should either lie flat on the floor with your helmet towards the area of detonation, or move away from any wall that might be penetrated by grenade fragments.
- ***High-Explosive, Dual-Purpose Grenade*--**The best round for engaging an urban threat is the M433 high-explosive, dual-purpose cartridge (TM 3-22.31 and Figure 8-11).

Chapter 8

Figure 8-11. M433 HEDP grenade.

Safety

8-20. It is easier to fire a grenade into an upper-story window using an M203 grenade launcher than it is to do throw it by hand.

- When someone must throw a hand grenade into an upper-story opening, he stands close to the building, using it for cover. He should only do this if the window opening has no glass or screening.
- He allows the grenade to cook off for at least 2 seconds, and then steps out far enough to lob the grenade into the upper-story opening. He keeps his weapon in his non-throwing hand, to use if needed. He *never* lays down his weapon, either outside or inside the building.
- The team must locate the nearest cover, in case the grenade falls back outside with them, instead of landing inside the building.
- Once a Soldier throws a grenade into the building, and it detonates, the team must move swiftly to enter the building or room.

CLEARING A ROOM

8-21. This paragraph discusses how to enter and clear a room:

SQUAD LEADER

1. Designates the assault team and identifies the location of the entry point for the team.
2. Positions the follow-on assault team to provide overwatch and supporting fires for the initial assault team.

ASSAULT TEAM

3. Moves as near the entry point as possible, using available cover and concealment.
4. If a supporting element is to perform an explosive or ballistic breach, remains in a covered position until after the breach. If necessary, provides overwatch and fire support for the breaching element.
5. Before moving to the entry point, team members signal each other that they are ready.
6. Avoids using verbal signals, which could alert the enemy.
7. To reduce exposure to fire, moves quickly from cover to the entry point.
8. Enters through the breach and, unless someone throws a grenade before the team enters, [the team] avoids stopping outside of the point of entry.

Urban Areas

TEAM LEADER (SOLDIER NO. 2)

9. Has the option of throwing a grenade into the room before entry. Grenade type (fragmentation, concussion, or stun type) depends on the ROE and the building structure.
10. If stealth is moot (not a factor), sounds off when he throws grenade, for example, *Frag out, Concussion out,* or *Stun out.*
11. If stealth is a factor, uses visual signals when he throws a grenade.

ASSAULT TEAM

12. On the signal to go, or immediately after the grenade detonates, moves through the entry point and quickly takes up positions inside the room. These positions must allow the team to completely dominate the room and eliminate the threat. Unless restricted or impeded, team members stop moving only after they clear the door and reach their designated point of domination. In addition to dominating the room, all team members identify possible loopholes and mouseholes in the ceiling, walls, and floor.

Note: Where enemy forces may be concentrated and the presence of noncombatants is unlikely, the assault team can precede their entry by throwing a fragmentation or concussion grenade (structure dependent) into the room, followed by aimed, automatic small-arms fire by the number-one Soldier as he enters.

SOLDIER NO. 1 (RIFLEMAN)

13. Enters the room and eliminates the immediate threat. Goes left or right, normally along the path of least resistance, toward one of two corners. When using a doorway as the point of entry, determines the path of least resistance based on the way the door opens.
    - If it opens outward, he moves toward the hinged side.
    - If it opens inward, he moves away from the hinges.
14. On entering, gauges the size of the room, the enemy situation, and any furniture or other obstacles to help him determine his direction of movement.

ASSAULT TEAM

15. Avoids planning where to move until the exact layout of the room is known. Then, each Soldier goes in the opposite direction from the Soldier in front of him. Every team member must know the sectors and duties of each position.

SOLDIER NO. 1

16. As the first Soldier goes through the entry point, he can usually see into the far corner of the room. He eliminates any immediate threat and, if possible, continues to move along the wall to the first corner. There he assumes a dominating position facing into the room.

TEAM LEADER (SOLDIER NO. 2)

17. Enters about the same time as Soldier No. 1, but as previously stated, moves in the opposite direction, following the wall and staying out of the center. He clears the entry point, the immediate threat area, and his corner, and then moves to a dominating position on his side of the room.

Chapter 8

GRENADIER (SOLDIER NO. 3)

18. Moves opposite Soldier No. 2 (team leader), at least 1 meter from the entry point, and then to a position that dominates his sector.

SAW GUNNER (SOLDIER NO. 4)

19. Moves opposite Soldier No. 3, and then to a position that dominates his sector.

> **Points of Domination**
> If the path of least resistance takes the first Soldier to the left, then all points of domination mirror those in the diagrams. Points of domination should be away from doors and windows to keep team members from silhouetting themselves.

ASSAULT TEAM

20. Ensures movement does not mask anyone's fire. On order, any member of the assault team may move deeper into the room, overwatched by the other team members. Once the team clears the room, the team leader signals to the squad leader that the room has been cleared. The squad leader marks the room IAW unit SOP. The squad leader determines whether his squad can continue to clear through the building. The squad reorganizes as necessary. Leaders redistribute the ammunition. The squad leader reports to the platoon leader when the room is clear.

Section III. FIGHTING POSITIONS

How do you find and use a fighting position properly? You have to know this: whether you are attacking or defending, your success depends on your ability to place accurate fire on the enemy--with the least exposure to return fire (Figure 8-12).

- Make maximum use of available cover and concealment.
- Avoid firing over cover; when possible, fire around it.
- Avoid silhouetting against light-colored buildings, the skyline, and so on.
- Carefully select a new fighting position before leaving an old one.
- Avoid setting a pattern. Fire from both barricaded and non-barricaded windows.
- Keep exposure time to a minimum.
- Begin improving your hasty position immediately after occupation.
- Use construction material that is readily available in an urban area.
- Remember: positions that provide cover at ground level may not provide cover on higher floors.

Figure 8-12. Some considerations for selecting and occupying individual fighting positions

Urban Areas

HASTY FIGHTING POSITION

8-22. A hasty fighting position is normally occupied in the attack or early stages of defense. It is a position from which you can place fire upon the enemy while using available cover for protection from return fire. You may occupy it voluntarily or be forced to occupy it due to enemy fire. In either case, the position lacks preparation before occupation. Some of the more common hasty fighting positions in an urban area are corners of buildings, behind walls, windows, unprepared loopholes, and the peak of a roof.

CORNERS OF BUILDINGS

8-23. You must be able to fire your weapon (both right and left-handed) to be effective around corners.

- A common error made in firing around corners is firing from the wrong shoulder. This exposes more of your body to return fire than necessary. By firing from the proper shoulder, you can reduce exposure to enemy fire (Figure 8-13).

Figure 8-13. Soldier firing left or right handed.

Chapter 8

- Another common mistake when firing around corners is firing from the standing position. If the Soldier exposes himself at the height the enemy expects, then he risks exposing the entire length of his body as a target for the enemy (Figure 8-14).

Figure 8-14. Soldier firing around a corner.

WALLS

8-24.   When firing from behind walls, you must fire around cover and not over it.

WINDOWS

8-25.   In an urban area, windows provide convenient firing ports. Avoid firing from the standing position, which would expose most of your body to return fire from the enemy, and which could silhouette you against a light-colored interior background. This is an obvious sign of your position, especially at night when the muzzle flash can be easily observed. To fire from a window properly, remain well back in the room to hide the flash, and kneel to limit exposure and avoid silhouetting yourself.

LOOPHOLES

8-26.   You may fire through a hole created in the wall and avoid windows. You must stay well back from the loophole so the muzzle of the weapon does not protrude beyond the wall, and the muzzle flash is concealed.

ROOF

8-27.   The peak of a roof provides a vantage point that increases field of vision and the ranges at which you can engage targets (Figure 8-15). A chimney, smokestack, or any other object protruding from the roof of a building should be used to reduce the size of the target exposed.

Urban Areas

Figure 8-15. Soldier firing from peak of a roof.

NO POSITION AVAILABLE

8-28. When subjected to enemy fire and none of the positions mentioned above are available, you must try to expose as little of yourself as possible. You can reduce your exposure to the enemy by lying in the prone position as close to a building as possible, on the same side of the open area as the enemy. In order to engage you, the enemy must then lean out the window and expose himself to return fire.

NO COVER AVAILABLE

8-29. When no cover is available, you can reduce your exposure by firing from the prone position, by firing from shadows, and by presenting no silhouette against buildings.

PREPARED FIGHTING POSITION

8-30. A prepared firing position is one built or improved to allow you to engage a particular area, avenue of approach, or enemy position, while reducing your exposure to return fire. Examples of prepared positions include barricaded windows, fortified loopholes, and sniper, antiarmor, and machine gun positions.

BARRICADED WINDOWS

8-31. The natural firing port provided by windows can be improved by barricading the window, leaving a small hole for you to use. Materials torn from the interior walls of the building or any other available material may be used for barricading.

8-32. Barricade all windows, whether you intend to use them as firing ports or not. Keep the enemy guessing. Avoid making neat, square, or rectangular holes, which clearly identify your firing positions to the enemy. For example, a barricaded window should not have a neat, regular firing port. The window should remain in its original condition so that your position is hard to detect. Firing from the bottom of the window gives you the advantage of the wall because the firing port is less obvious to the enemy. Sandbags are used to reinforce the wall below the window and to increase protection. All glass must be removed from the window to prevent injury. Lace curtains permit you to see out and prevent the enemy from seeing in. Wet blankets should be placed under weapons to reduce dust. Wire mesh over the window keeps the enemy from throwing in hand grenades.

Chapter 8

LOOPHOLES

8-33. Although windows usually are good fighting positions, they do not always allow you to engage targets in your sector. To avoid establishing a pattern of always firing from windows, alternate positions, for example, fire through a rubbled outer wall, from an interior room, or from a prepared loophole. The prepared loophole involves cutting or blowing a small hole into the wall to allow you to observe and engage targets in your sector. Use sandbags to reinforce the walls below, around, and above the loophole.

Protection--Two layers of sandbags are placed on the floor to protect you from an explosion on a lower floor (if the position is on the second floor or higher). Construct a wall of sandbags, rubble, and furniture to the rear of the position as protection from explosions in the room. A table, bedstead, or other available material can provide OHC for the position. This cover prevents injury from falling debris or explosions above your position.

Camouflage--Hide the position in plain sight by knocking other holes in the wall, making it difficult for the enemy to determine which hole the fire is coming from. Remove exterior siding in several places to make loopholes less noticeable. Due to the angled firing position associated with loopholes, you can use the same loophole for both primary and supplementary positions. This allows you to shift your fire easily onto a sector that was not previously covered by small arms fire.

Backblast--SLM and CCMs crews may be hampered in choosing firing positions due to the backblast of their weapons. They may not have enough time to knock out walls in buildings and clear backblast areas. They should select positions that allow the backblast to escape such as corner windows where the round fired goes out one window and the backblast escapes from another. When conducting defensive operations the corner of a building can be improved with sandbags to create a firing position.

Shoulder-Launched Munitions and Close Combat Missiles--Various principles of employing SLM and CCMs weapons have universal applications. These include using available cover, providing mutual support, and allowing for backblast. However, urban areas require additional considerations. Soldiers must select numerous alternate positions, particularly in structures without cover from small-arms fire. Soldiers must position their weapons in the shadows and within the building.

-- A gunner firing an AT4 or Javelin from the top of a building can use a chimney for cover, if available. He should reinforce his position by placing sandbags to the rear so they do not interfere with the backblast.

-- When selecting firing positions for his SLM or CCM, he uses rubble, corners of buildings, or destroyed vehicles as cover. He moves his weapon along rooftops to find better angles for engaging enemy vehicles. On tall buildings, he can use the building itself as overhead cover. He must select a position where backblast will not damage or collapse the building, or injure him.

> **DANGER**
>
> When firing within an enclosure, ensure that it measures at least 10 feet by 15 feet (150 square feet); is clear of debris and other loose objects; and has windows, doors, or holes in the walls where the backblast can escape.

-- The machine gunner can emplace his weapon almost anywhere. In the attack, windows and doors offer ready-made firing ports (Figure 8-16). For this reason, avoid windows and doors, which the enemy normally has under observation and fire. Use any opening created in walls during the fighting. Small explosive charges can create loopholes for machine gun positions. Regardless of the openings used, ensure machine guns are inside the building and that they remain in the shadows.

Urban Areas

Figure 8-16. Emplacement of machine gun in a doorway.

-- Upon occupying a building, board up all windows and doors. Leave small gaps between the boards so you can use windows and doors as alternate positions.
-- Use loopholes extensively in the defense. Avoid constructing them in any logical pattern, or all at floor or tabletop levels. Varying height and location makes them hard to pinpoint and identify. Make dummy loopholes and knock off shingles to aid in the deception. Construct loopholes behind shrubbery, under doorjambs, and under the eaves of a building, because these are hard to detect. In the defense, as in the offense, you can construct a firing position to use the building for OHC.
-- You can increase your fields of fire by locating the machine gun in the corner of the building or in the cellar. To add cover and concealment, integrate available materials, such as desks, overstuffed chairs, couches, and other items of furniture, into the construction of bunkers.
-- Grazing fire is ideal, but sometimes impractical or impossible. Where destroyed vehicles, rubble, and other obstructions restrict the fields of grazing fire, elevate the gun to allow you to fire over obstacles. You might have to fire from second or third story loopholes. You can build a firing platform under the roof, and then construct a loophole. Again, conceal the exact location of the position. Camouflage the position by removing patches of shingles over the entire roof.

This page intentionally left blank.

Chapter 9
'Every Soldier is a Sensor'

Every Soldier, as a part of a small unit, can provide useful information and is an essential component to the commanders achieving situational understanding. This task is critical, because the environment in which Soldiers operate is characterized by violence, uncertainty, complexity, and asymmetric methods by the enemy. The increased situational awareness that you must develop through personal contact and observation is a critical element of the friendly force's ability to more fully understand the operational environment. Your life and the lives of your fellow Soldiers could depend on reporting what you see, hear, and smell.

DEFINITION

9-1. The 'Every Soldier is a Sensor' (ES2) concept ensures that Soldiers are trained to actively observe for details for the commander's critical information requirement (CCIR) while in an AO. It also ensures they can provide concise, accurate reports. Leaders will know how to collect, process, and disseminate information in their unit to generate timely intelligence. They should establish a regular feedback and assessment mechanism for improvement in implementing ES2. Every Soldier develops a special level of exposure to events occurring in the AO and can collect information by observing and interacting with the environment. Intelligence collection and development is everyone's responsibility. Leaders and Soldiers should fight for knowledge in order to gain and maintain greater situational understanding.

RESOURCES

9-2. As Soldiers develop the special level of exposure to the events occurring in their operating environment, they should keep in mind certain potential indicators as shown in Figure 9-1, page 9-2. These indicators are information on the intention or capability of a potential enemy that commanders need to make decisions. You will serve as the commander's "eyes and ears" when--

- Performing traditional offensive or defensive missions.
- Patrolling in a stability and reconstruction or civil support operation.
- Manning a checkpoint or a roadblock.
- Occupying an observation post.
- Passing through areas in convoys.
- Observing and reporting elements of the environment.
- Observing and reporting activities of the populace in the area of operations.

SIGHT	SOUND	TOUCH	SMELL
Look for--	Listen for--	Feel for--	Smell for--
• Enemy personnel, vehicles, and aircraft • Sudden or unusual movement • New local inhabitants • Smoke or dust • Unusual movement of farm or wild animals • Unusual activity--or lack of activity--by local inhabitants, especially at times or places that are normally inactive or active • Vehicle or personnel tracks • Movement of local inhabitants along uncleared routes, areas, or paths • Signs that the enemy has occupied the area • Evidence of changing trends in threats • Recently cut foliage • Muzzle flashes, lights, fires, or reflections • Unusual amount (too much or too little) of trash	• Running engines or track sounds • Voices • Metallic sounds • Gunfire, by weapon type • Unusual calm or silence • Dismounted movement • Aircraft	• Warm coals and other materials in a fire • Fresh tracks • Age of food or trash	• Vehicle exhaust • Burning petroleum products • Food cooking • Aged food in trash • Human waste
	OTHER CONSIDERATIONS		
	Armed Elements	Locations of factional forces, mine fields, and potential threats.	
	Homes and Buildings	Condition of roofs, doors, windows, lights, power lines, water, sanitation, roads, bridges, crops, and livestock.	
	Infrastructure	Functioning stores, service stations, and so on.	
	People	Numbers, gender, age, residence or DPRE status, apparent health, clothing, daily activities, and leadership.	
	Contrast	Has anything changed? For example, are there new locks on buildings? Are windows boarded up or previously boarded up windows now open, indicating a change in how a building is expected to be used? Have buildings been defaced with graffiti?	

Figure 9-1. Potential indicators.

9-3. Commanders get information from many sources, but you are his best source. You can in turn collect information from the following sources:

- Enemy prisoners of war (EPWs)/detainees are an immediate source of information. Turn captured Soldiers over to your leader quickly. Also, tell him anything you learn from them.
- Captured enemy documents (CEDs) may contain valuable information about present or future enemy operations. Give such documents to your leader quickly.
- Captured enemy equipment (CEEs) eliminates an immediate threat. Give such equipment to your leader quickly.
- Enemy activity (the things the enemy is doing) often indicates what the enemy plans to do. Report everything you see the enemy do. Some things that may not seem important to you may be important to your commander.
- Tactical questioning, observation, and interaction with displaced persons, refugees, or evacuees (DPRE), during the conduct of missions, can yield important information.
- Local civilians, however often have the most information about the enemy, terrain, and weather in a particular area. Report any information gained from civilians. However, you cannot be sure

which side the civilians are trying to help, so be careful when acting on information obtained from them. If possible, try to confirm the information by some other means.

FORMS OF QUESTIONING

9-4. Questioning may be achieved by tactical or direct methods. The following paragraphs detail both methods:

Tactical Questioning--Tactical questioning is the initial questioning for information of immediate value. When the term applies to the interaction with the local population, it is not really questioning but is more conversational in nature. The task can be designed to build rapport as much, and collect information and understand the environment. You will conduct tactical questioning based on your unit is SOPs, ROE, and the order for that mission. Your leaders must include specific guidance for tactical questioning in the operation order (OPORD) for appropriate missions. Information reported because of tactical questioning is passed up through your chain of command to the battalion/brigade intelligence officers (S-2) and assistant chief of staff for intelligence (G-2), which forms a vital part of future planning and operations. Additionally, you are not allowed to attempt any interrogation approach techniques in the course of tactical questioning.

Direct Questioning--Direct questioning is an efficient method of asking precise questions according to a standard pattern. The goal is to obtain the maximum amount of intelligence information in the least amount of time. Direct questions must clearly indicate the topic being questioned as they require an effective narrative response (i.e., be brief, simple, but specific). Clearly define each subject using a logical sequence. Basic questions are used to discourage "yes" or "no" answers. Direct questioning is the only technique authorized for ES2 tactical questioning. Soldiers who are not trained and certified interrogators are forbidden to attempt to apply any interrogation approach techniques. When it is clear that the person being questioned has no further information, or does not wish to cooperate further, tactical questioning must stop.

9-5. Various AOs will have different social and regional considerations that can affect communications and the conduct of operations (i.e., social behaviors, customs, and courtesies). You must also be aware of the following safety and cultural considerations:

- Know the threat level and force protection (FP) measures in your AO.
- Know local customs and courtesies.
- Avoid using body language that locals might find rude.
- Approach people in normal surroundings to avoid suspicion.
- Behave in a friendly and polite manner.
- Remove sunglasses when speaking to those people with whom you are trying to create a favorable impression.
- Know as much as possible about the local culture, including a few phrases in the local language.
- If security conditions permit, position your weapon in the least intimidating position as possible.

REPORT LEVELS

9-6. All information collected by patrols, or via other contact with the local population, is reported through your chain of command to the unit S-2. The S-2 is responsible for transmitting the information through intelligence channels to the supported military intelligence elements, according to unit intelligence tasks and the OPORD for the current mission. Therefore, if everyone is involved in the collection of combat information, then everyone must be aware of the priority intelligence requirements (PIR). All Soldiers who have contact with the local population and routinely travel within the area must know the CCIR, and their responsibility to observe and report. The four levels of mission reports follow:

Chapter 9

LEVEL 1

9-7. Information of critical tactical value is reported immediately to the S-2 section, while you are still out on patrol. These reports are sent via channels prescribed in the unit SOP. The size, activity, location, uniform, time, equipment (SALUTE) format is an example of Level I reporting.

LEVEL 2

9-8. Immediately upon return to base, the patrol will conduct an after action review (AAR) and write a patrol report. The format may be modified to more thoroughly capture mission-specific information. This report is passed along to the S-2 section prior to a formal debriefing. Your leaders must report as completely and accurately as possible since this report will form the basis of the debriefing by the S-2 section.

LEVEL 3

9-9. After receiving the initial patrol report, the S-2 section will debrief your patrol for further details and address PIR and CCIR not already covered in the patrol report.

LEVEL 4

9-10. Follow-up reporting is submitted as needed after the unit S-2 section performs the debriefing.

Note: Any patrols or activities should be preceded by a prebriefing, which is a consolidated summary of the AOs historical activities.

SALUTE FORMAT

9-11. These four levels help the unit S-2 section record and disseminate both important and subtle details of for use in all-source analysis, future planning, and passing on to higher S-2/G-2. This information helps them analyze a broad range of information and disseminate it back to your level and higher. Report all information about the enemy to your leader quickly, accurately, and completely. Such reports should answer the questions *who*, *what*, and *where* after *when*. Use the SALUTE format when reporting. Make notes and draw sketches to help you remember details. Table 9-1 shows how to use the SALUTE format.

'Every Soldier is a Sensor'

Table 9-1. SALUTE format line by line.

Line No.	Type Info	Description
1	(S)ize/Who	Expressed as a quantity and echelon or size. For example, report "10 enemy Infantrymen" (not "a rifle squad").
		If multiple units are involved in the activity you are reporting, you can make multiple entries.
2	(A)ctivity/What	Relate this line to the PIR being reported. Make it a concise bullet statement. Report what you saw the enemy doing, for example, "emplacing mines in the road."
3	(L)ocation/Where	This is generally a grid coordinate, and should include the 100,000-meter grid zone designator. The entry can also be an address, if appropriate, but still should include an eight-digit grid coordinate. If the reported activity involves movement, for example, advance or withdrawal, then the entry for location will include "from" and "to" entries. The route used goes under "Equipment/How."
4	(U)nit/Who	Identify who is performing the activity described in the "Activity/What" entry. Include the complete designation of a military unit, and give the name and other identifying information or features of civilians or insurgent groups.
5	(T)ime/When	For future events, give the DTG for when the activity will initiate. Report ongoing events as such. Report the time you saw the enemy activity, not the time you report it. Always report local or Zulu (Z) time.
6	(E)quipment/How	Clarify, complete, and expand on previous entries. Include information about equipment involved, tactics used, and any other essential elements of information (EEI) not already reported in the previous lines.

HANDLING AND REPORTING OF THE ENEMY

9-12. The following paragraphs detail adequate protocol for handling enemy documents, EPWs, and equipment:

CAPTURED ENEMY DOCUMENTS

9-13. A CED is defined as any piece of recorded information obtained from the threat. CEDs are generally created by the enemy, but they can also be US or multinational forces documents that were once in the hands of the enemy. CEDs can provide crucial information related to answering the commander's PIR or even be exploited to put together smaller pieces of an overall situation.

9-14. Every confiscated or impounded CED must be tagged and logged before being transferred through the appropriate channels. The tag contains the specifics of the item, and the log is a simple transmittal document used to track the transfer of CEDs between elements. Your leaders are responsible for creating the initial CED log.

9-15. While the information required is formatted, any durable field expedient material can be used as a CED tag if an official tag is unavailable. Ensure that the writing is protected from the elements by covering it with plastic or transparent tape. The importance of the tag is that it is complete and attached to the CED it represents. The following information, at a minimum, should be recorded on a CED tag. Instructions for filling out the tag follow (Figure 9-2):

Nationality--Detail the country of origin of the unit that captured the enemy document.

Date-Time Group--Include date and time of capture.

Place--Include a six-to eight-digit grid coordinate and describe the location where the document was captured.

Identity--Define where the CED came from, its owner, and so on.

Circumstances--Describe how the CED was obtained.

Description--Briefly describe the CED. Enough information should be annotated for quick recognition.

```
┌─────────────────────────────────────────────┐
│           CAPTURED DOCUMENT TAG             │
│                                             │
│  NATIONALITY OF CAPTURING FORCE _____ │
│                                             │
│  DATE/TIME CAPTURED _____ │
│                                             │
│  PLACE CAPTURED _____ │
│                                             │
│  CAPTURING UNIT _____ │
│                                             │
│  IDENTITY OF SOURCE (If Applicable) _____ │
│                                             │
│  CIRCUMSTANCES OF CAPTURE _____ │
│                                             │
│  DESCRIPTION OF WEAPON/DOCUMENT _____ │
└─────────────────────────────────────────────┘
```

Figure 9-2. Example captured document tag.

TREATMENT OF EPWS AND DETAINEES

9-16. EPWs/detainees are a good source of information. They must be handled without breaking international law and without losing a chance to gain intelligence. Treat EPWs humanely. Do not harm them, either physically or mentally. The senior Soldier present is responsible for their care. If EPWs cannot be evacuated in a reasonable time, give them food, water, and first aid. Do not give them cigarettes, candy, or other comfort items. EPWs who receive favors or are mistreated are poor interrogation subjects. In handling EPWs/detainees, follow the procedure of search, segregate, silence, speed, safeguard, and tag (the 5 Ss and T). It implies the legal obligation that each Soldier has to treat an individual in custody of, or under the protection of, US Soldiers humanely. The 5 Ss and T are conducted as follows:

Search--This indicates a thorough search of the person for weapons and documents. You must search and record the EPWs/detainees equipment and documents separately. Record the description of weapons, special equipment, documents, identification cards, and personal affects on the capture tag.

Silence--Do not allow the EPWs/detainees to communicate with one another, either verbally or with gestures. Keep an eye open for potential troublemakers, both talkers or quiet types, and be prepared to separate them.

Segregate--Keep civilians and military separate, and then further divide them by rank, gender, nationality, ethnicity, and religion. This technique helps keep them quiet.

Safeguard--Provide security for and protect the EPWs/detainees. Get them out of immediate danger and allow them to keep their personal chemical protective gear, if they have any, and their identification cards.

Speed--Information is time sensitive. It is very important to move personnel to the rear as quickly as possible. The other thing to consider is that an EPW/detainee's resistance to questioning grows as time goes on. The initial shock of being captured or detained wears off and they begin to think of escape.

Note: Exercising speed, in this instance, is critical because the value of information erodes in a few hours. Human intelligence (HUMINT) Soldiers who are trained and who have the appropriate time and means will be waiting to screen and interrogate these individuals.

PERSONNEL AND EQUIPMENT TAGS

9-17. Use wire, string, or other durable material to attach Part A, DD Form 2745, *Enemy Prisoner of War (EPW) Capture Tag*, or a field expedient alternative, to the detainee's clothing. Tell him not to remove or alter the tag. Attach another tag to any confiscated property. On each tag, write the following, making sure that your notes clearly link the property with the person from whom you confiscated it:

- Date and time of capture.
- Location of the capture (grid coordinates).
- Capturing unit.
- Circumstances of capture (why person was detained).
  - -- Who?
  - -- What?
  - -- Where?
  - -- Why?
  - -- Witnesses?

OPERATIONS SECURITY

9-18. Operations security (OPSEC) is the process your leaders follow to identify and protect essential elements of friendly information (EEFI). The Army defines EEFI as critical aspects of a friendly operation that, if known by the enemy, would subsequently compromise, lead to failure, or limits success of the operation and therefore must be protected from detection. All Soldiers execute OPSEC measures as part of FP. Effective OPSEC involves telling Soldiers exactly why OPSEC measures are important, and what they are supposed to accomplish. You must understand that the cost of failing to maintain effective OPSEC can result in the loss of lives. Understanding why you are doing something and what your actions are supposed to accomplish, allows you and your fellow Soldiers to execute tasks more effectively. However, this means that you and your fellow Soldiers must--

- Avoid taking personal letters or pictures into combat areas.
- Avoid keeping diaries in combat areas.
- Practice camouflage principles and techniques.
- Practice noise and light discipline.
- Practice field sanitation.
- Use proper radiotelephone procedure.
- Use the challenge and password properly.
- Abide by the Code of Conduct (if captured).
- Report any Soldier or civilian who is believed to be serving with or sympathetic to the enemy.
- Report anyone who tries to get information about US operations.

Chapter 9

- Destroy all maps or important documents if capture is imminent.
- Avoid discussing military operations in public areas.
- Discuss military operations only with those persons having a need to know the information.
- Remind fellow Soldiers of their OPSEC responsibilities.

OBSERVATION TECHNIQUES

9-19. During all types of operations, you will be looking for the enemy. However, there will be times when you will be posted in an OP to watch for enemy activity. An OP is a position from which you watch an assigned sector of observation and report all activity seen or heard in your sector.

DAY OBSERVATION

9-20. In daylight, use the visual search technique to search terrain. You must visually locate and distinguish enemy activity from the surrounding terrain features by using the following scanning techniques:

Rapid Scan--This is used to detect obvious signs of enemy activity. It is usually the first method you will use (Figure 9-3). To conduct a rapid scan--
- Search a strip of terrain about 100 meters deep, from left-to-right, pausing at short intervals.
- Search another 100-meter strip farther out, from right-to-left, overlapping the first strip scanned, pausing at short intervals.
- Continue this method until the entire sector of fire has been searched.

Slow Scan--The slow scan search technique uses the same process as the rapid scan but much more deliberately, which means a slower, side-to-side movement and more frequent pauses (Figure 9-5).

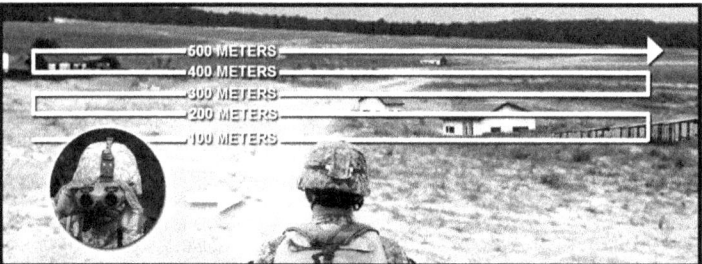

Figure 9-3. Rapid/slow-scan pattern.

Detailed Search--If you find no targets using either the rapid or slow scan techniques, make a careful, detailed search of the target area using M22 binoculars. The detailed search is like the slow scan, but searching smaller areas with frequent pauses and almost incremental movement. The detailed search, even more than the rapid or slow scan, depends on breaking a larger sector into smaller sectors to ensure everything is covered in detail and no possible enemy positions are overlooked (Figure 9-4). You must pay attention to the following:
-- Likely enemy positions and suspected vehicle/dismounted avenues of approach.
-- Target signatures, such as road junctions, hills, and lone buildings, located near prominent terrain features.
-- Areas with cover and concealment, such as tree lines and draws.

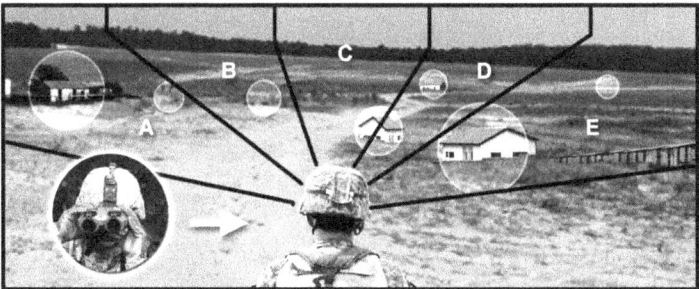

Figure 9-4. Detailed search.

LIMITED VISIBILITY OBSERVATION

9-21. Although operating at night has definite advantages, it is also difficult. Your eyes do not work as well as during the day, yet they are crucial to your performance. You need to be aware of constraints your eyes place upon you at night, because 80 percent of your sensory input comes through them. Your ability to see crisp and clear images is significantly reduced.

Dark Adaptation

9-22. Dark adaptation is the process by which the human body increases the eye's sensitivity to low levels of light. Adaptation to darkness occurs at varying degrees and rates. During the first 30 minutes in the dark, eye sensitivity increases about 10,000 times. Dark adaptation is affected by exposure to bright light such as matches, flashlights, flares, or vehicle headlights. Full recovery from these exposures can take up to 45 minutes. Your color perception decreases at night. You may be able to distinguish light and dark colors depending on the intensity of reflected light. At night, bright warm colors such as reds and oranges are hard to see and will appear dark. In fact, reds are nearly invisible at night. Unless a dark color is bordered by two lighter colors, it is invisible. On the other hand, greens and blues will appear brighter, although you may not be able to determine their color. Since visual sharpness at night is one-seventh of what it is during the day, you can see only large, bulky objects, so you must recognize objects by their general shape or outline. Knowing the design of structures common in the AO will help you determine shape or silhouette. Darkness also reduces depth perception.

- *Normal Blind Spots*--The normal blind spot is always present, day and night. It is caused by the lack of light receptors where the optic nerve inserts into the back of the eye. The normal blind spot occurs when you use just one eye. When you close the other eye, objects about 12 to 15 degrees away from where you are looking will disappear. When you uncover your eye, the objects will reappear.
- *Night Blind Spots*--When you stare at an object at night, under starlight or lower levels of illumination, it can disappear or fade away. This is a result of the night blind spot. It affects both eyes at the same time and occurs when using the central vision of both eyes. Consequently, larger and larger objects are missed as the distances increase. In order to avoid the night blind spots, look to all sides of objects you are trying to find or follow. Do not stare. This is the only way to maximize your night vision.

Night Observation Techniques

9-23. The following paragraphs detail night observation techniques:

Dark Adaptation Technique--First, let your eyes become adjusted to the darkness. Do so by staying either in a dark area for about 30 minutes, or in a red-light area for about 20 minutes followed by about 10 minutes in a dark area. The red-light method may save time by allowing you to get orders, check equipment, or do some other job before moving into darkness.

Night Vision Scans--Dark adaptation is only the first step toward making the greatest use of night vision. Scanning enables you to overcome many of the physiological limitations of your eyes (Figure 9-5). It can also reduce confusing visual illusions or your eyes playing tricks on you. This technique involves looking from right to left or left to right using a slow, regular scanning movement. At night, it is essential to avoid looking directly at a faintly visible object when trying to confirm its presence.

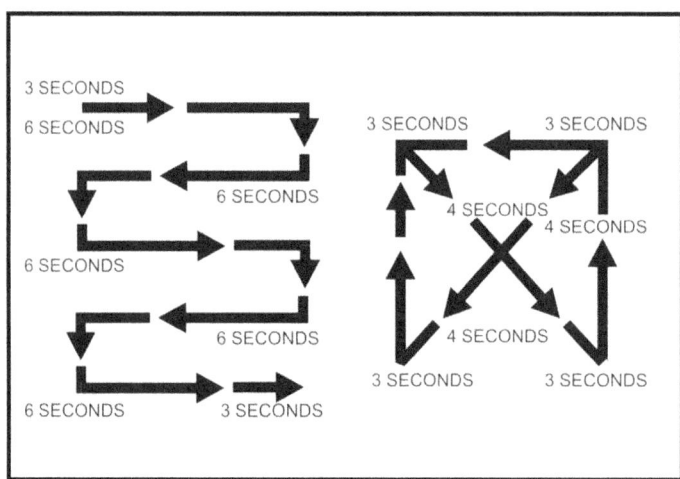

Figure 9-5. Typical scanning pattern.

Off-Center Vision--The technique of viewing an object using central vision is ineffective at night. Again, this is due to the night blind spot that exists during low illumination (Figure 9-6). You must learn to use off-center vision. This technique requires viewing an object by looking 10 degrees above, below, or to either side of it rather than directly at it. Additionally, diamond viewing is very similar in that you move your eyes just slightly, a few degrees, in a diamond pattern around the object you wish to see. However, the image of an object bleaches out and becomes a solid tone when viewed longer than 2 or 3 seconds. You do not have to move your head to use your peripheral vision. By shifting your eyes from one off-center point to another, you can continue to pick-up the object in your peripheral field of vision.

'Every Soldier is a Sensor'

Figure 9-6. Off-center viewing.

LIMITED VISIBILITY DEVICES

9-24. The three devices used to increase lethality at night include night vision devices (NVDs), thermal weapon sights, and aiming lasers. Each provides different views of the infrared (IR) spectrum, which is simple energy. The electromagnetic spectrum is simply energy (light). Before you can fully operate these devices, you must know how they work in the IR range, and you must know the electromagnetic (light) spectrum. You should also know the advantages and disadvantages of each piece of equipment. This is the only way to know when to employ which.

Image-Intensification Devices--An image intensifier captures ambient light, and then amplifies it thousands of times electronically, allowing you to see the battlefield through *night vision goggles* (NVGs). Ambient light comes from the stars, moon, or sky glow from distant man-made sources such as cities. Humans can only see part of this spectrum of light with the naked eye. Just beyond red visible light is infrared (IR) light, which is broken down into three ranges--near, middle, and far infrared. Leaders can conduct combat missions with no active illumination sources, just image intensifiers. However, the main advantages of image intensifiers as NVDs are their small sizes, light weights, and low power requirements. Image intensifiers increase vision into the IR range. They rely on ambient light and energy in the near IR range. This energy emits from natural and artificial sources such as moonlight, starlight, and city lights. Image intensifiers include the following (Figure 9-7):
-- AN/PVS-7A/B/C/D.
-- AN/PVS-14.

Chapter 9

Figure 9-7. AN/PVS-7 and AN/PVS-14.

Thermal Imaging Devices

9-25. The second type of device that uses IR light is the thermal imaging device (Figure 9-8). This type of device detects electromagnetic radiation (heat) from humans and man-made objects, and translates that heat into an electronic image. Thermal imagers operate the same regardless of the level of ambient light. Thermal weapon sights (TWSs) operate in the middle to far IR ranges. These sights detect IR light emitted from friction, from combustion, or from any objects that are radiating natural thermal energy. Since the TWS and other thermal devices operate within the middle/far IR range, they cannot be used with image intensifiers. Thermal devices can be mounted on a weapon or handheld. The TWS works well day or night. It has excellent target acquisition capabilities, even through fog, haze, and conventional battlefield smoke.

- AN/PAS-13(V1) light weapon thermal sight (LWTS).
  -- M16- and M4-series rifles and carbines
  -- M136 (AT4) light antiarmor weapon
- AN/PAS-13(V2) medium weapon thermal sight (MWTS)
  -- M249 machine gun
  -- M240B series medium machine gun
- AN/PAS-13(V3) heavy weapon thermal sight (HWTS)
  -- M24 Sniper rifle
  -- M107 Sniper rifle
  -- M2 (50 Cal.) HB machine gun
  -- MK 19 machine gun

'Every Soldier is a Sensor'

Figure 9-8. AN/PAS-13, V1, V2, and V3.

Aiming Lasers

9-26. Aiming lasers--both the AN/PAQ-4-series and the AN/PEQ-2A (Figure 9-9)--also operate in the electromagnetic spectrum, specifically in the near IR range. [These lasers] are seen through image-intensification devices. The aiming lasers cannot be used in conjunction with the TWS, because the latter operates in the middle to far IR spectrum.

Figure 9-9. AN/PAQ-4-series and the AN/PEQ-2A.

Chapter 9

PROPER ADJUSTMENTS TO THE IMAGE INTENSIFIERS

9-27. You must make the proper adjustments to the image intensifiers in order to get the best possible picture. The aiming lasers cannot be seen with the unaided eye; they can only be seen with image intensification devices. You must know how these devices work to maximize the quality of what is being viewed by making the proper adjustments to these devices.

Scanning

9-28. The NVDs have a 40-degree field of view (FOV) leaving the average shooter to miss easy targets of opportunity, more commonly the 50-meter left or right target. You must train to aggressively scan your sector of fire for targets. Target detection at night is only as good as you practice. Regular blinking during scanning, which must be reinforced during training, relieves some of the eyestrain from trying to spot far targets. After you have mastered the art of scanning, you will find that targets are easier to detect by acknowledging the flicker or movement of a target.

Walking

9-29. Once a target has been located, you must be aware of the placement of the aiming laser. Laser awareness is necessary. If you activate your laser and it is pointing over the target into the sky, you will waste valuable time trying to locate exactly where your laser is pointing. Also, it increases your chances of being detected and fired upon by the enemy. When engaging a target, aim the laser at the ground just in front of the target, walk the aiming laser along the ground and up the target until you are center mass, and then engage the target. Walking your laser to the target is a quick and operationally secure means of engaging the enemy with your aiming laser.

IR Discipline

9-30. Once a target has been located and engaged with the aiming laser, the laser must be deactivated. On the range, IR discipline means actively scanning with the laser off. Once a target is located, walk the laser to the target and engage. After the target has been engaged, the laser goes off.

RANGE ESTIMATION

9-31. You must often estimate ranges. You must accurately determine distance and prepare topographical sketches or range cards. Your estimates will be easier to make and more accurate if you know various range-estimation techniques.

FACTORS

9-32. Three factors affect range estimates:

Nature of the Object

Outline	An object of regular outline, such as a house, appears closer than one of irregular outline, such as a clump of trees.
Contrast	A target that contrasts with its background appears to be closer than it actually is.
Exposure	A partly exposed target appears more distant than it actually is.

Nature of Terrain

Contoured terrain	Looking across contoured terrain makes an object seem farther.
Smooth terrain	Looking across smooth terrain, such as sand, water, or snow, makes a distant object seem nearer.
Downhill	Looking downhill at an object makes it seem farther.
Uphill	Looking uphill at an object makes it seem nearer.

Light Conditions

Sun behind observer	A front-lit object seems nearer.
Sun behind object	A back-lit object seems farther away.

ESTIMATION METHODS

9-33. Methods of range estimation include--

- The 100-meter unit-of-measure method.
- The appearance-of-objects method.
- The flash-and-sound method.
- The mil-relation method.
- A combination of these.

100-Meter-Unit-of-Measure Method--

9-34. Picture a distance of 100 meters on the ground. For ranges up to 500 meters, count the number of 100-meter lengths between the two points you want to measure. Beyond 500 meters, pick a point halfway to the target, count the number of 100-meter lengths to the halfway point, and then double that number to get the range to the target. The accuracy of the 100-meter method depends on how much ground is visible. This is most true at long ranges. If a target is at a range of 500 meters or more, and you can only see part of the ground between yourself and the target, it is hard to use this method with accuracy. If you know the apparent size and detail of troops and equipment at known ranges, then you can compare those characteristics to similar objects at unknown ranges. When the characteristics match, the range does also.

Appearance-of-Object Method

9-35. To use the appearance-of-objects method, you must be familiar with characteristic details of objects as they appear at various ranges. As you must be able to see those details to make the method work, anything that limits visibility (such as weather, smoke, or darkness) will limit the effectiveness of this method. If you know the apparent size and detail of troops and equipment at known ranges, then you can compare those characteristics to similar objects at unknown ranges. When the characteristics match, the range does also. Table 9-2 shows what is visible on the human body at specific ranges.

Chapter 9

Table 9-2. Appearance of a body using appearance-of-objects method.

RANGE (in meters)	WHAT YOU SEE
200	Clear in all detail such as equipment, skin color
300	Clear body outline, face color good, remaining detail blurred
400	Body outline clear, other details blurred
500	Body tapered, head indistinct from body
600	Body a wedge shape, with no head apparent
700	Solid wedge shape (body outline)

Flash-and-Sound Method

9-36. This method is best at night. Sound travels through air at 1,100 feet (300 meters) per second. That makes it possible to estimate distance if you can both see and hear a sound-producing action. When you see the flash or smoke of a weapon, or the dust it raises, immediately start counting. Stop counting when you hear the sound associated with the action. The number at which you stop should be multiplied by three. This gives you the approximate distance to the weapon in hundreds of meters. If you stop at one, the distance is about 300 meters. If you stop at three, the distance is about 900 meters. When you must count higher than nine, start over with one each time you hit nine. Counting higher numbers throws the timing off.

Mil-Relation Formula

9-37. This is the easiest and best way to estimate range. At 1,000 meters, a 1-mil angle equals 1 meter (wide or high). To estimate the range to a target, divide the estimated height of the target in meters (obtained using the reticle in the M22 binoculars) by the size of the target in mils. Multiply by 1,000 to get the range in meters (Figure 9-10).

Figure 9-10. Mil-relation formula.

Combination of Methods

9-38. Battlefield conditions are not always ideal for estimating ranges. If the terrain limits the use of the 100-meter unit-of-measure method, and poor visibility limits the use of the appearance-of-objects method, you may have to use a combination of methods. For example, if you cannot see all of the terrain out to the target, you can still estimate distance from the apparent size and detail of the target itself. A haze may obscure the target details, but you may still be able to judge its size or use the 100-meter method. By using either one or both of the methods, you should arrive at a figure close to the true range.

Chapter 10
Combat Marksmanship

Combat marksmanship is essential to all Soldiers—not only to acquire the expert skills necessary for survival on the battlefield, but, because it enforces teamwork and discipline. In every organization, all members must continue to practice certain skills to remain proficient. Marksmanship is paramount.

This chapter discusses several aspects of combat marksmanship, including safety, administrative, weapons.

SAFETY

This paragraph describes procedures and requirements for handling all organic and special weapons. These procedures are designed to prevent safety-related accidents and fratricide. They are intended for use in both training and combat, and apply to all assigned weapons of a unit. In all cases, strict self-discipline is the most critical factor for the safe handling of weapons. The procedures for weapons handling may vary based on METT-TC, but the following procedures are generally recommended:

- As soon as you are issued a weapon, immediately clear it, and place it on safe IAW the appropriate Soldier's or operator's manual.
- Keep the weapon on safe, except--
  -- When it is stored in an arms room.
  -- Immediately before target engagement.
  -- When directed by the chain of command.
- At all times, handle weapons as if they are loaded.
- Never point a weapon at anyone, unless a life-threatening situation justifies the use of deadly force.
- Always know where you are pointing the muzzle of the weapon.
- Always know if the weapon is loaded.
- Always know if the weapon is on safe.
- Insert magazines or belts of ammunition only on the direction of your chain of command.

Note: The chain of command determines when to load the weapon and chamber a round in reference to METT-TC in a combat environment. Generally, weapons remain on safe until ready to fire.

ADMINISTRATIVE PROCEDURES

10-1. Administrative procedures include weapons clearing, grounding of weapons, and aircraft and vehicle movement. Soldiers must know how to handle their weapons and have a clear understanding of fire control.

Chapter 10

ADMINISTRATIVE WEAPONS CLEARING

10-2. Administrative weapons clearing is performed following the completion of the tactical phase of all live fires and range qualifications, or upon reentry of a secure area in combat. Magazines or belts are removed from all weapons. The chain of command inspects all chambers visually, using red filtered light if at night, and verifies that each weapon and magazine is clear of ammunition. Weapons should also be rodded. Magazines are not reinserted into weapons.

GROUNDING OF WEAPONS

10-3. If grounded with equipment, all weapons are placed on SAFE and arranged off the ground with the open chamber visible, if applicable. Bipod-mounted weapons are grounded on bipods with all muzzles facing downrange and away from nearby Soldiers.

AIRCRAFT AND VEHICLE MOVEMENT

10-4. Weapons should always be cleared and on SAFE when conducting movement in aircraft and vehicles, unless leaders issue specific instructions to do otherwise. Keep weapon muzzles pointed downward when traveling on aircraft. Weapons are locked and loaded only after exiting the aircraft or vehicle, or upon command of the leader.

> *Note:* Take extra care to correctly handle the pistol, especially the M9 with its double action (fire from the hammer down) feature. Removing the pistol from the holster can accidentally move the safety lever to fire, permitting immediate double action mode of fire. Only chamber a round in a pistol when a specific threat warrants doing so.

WEAPONS

10-5. Weapons include the M9 pistol; M16-series rifles; M4 carbine rifles; M203 grenade launcher; M249 squad automatic weapon (SAW); M240B machine gun; M2 .50 caliber machine gun; MK 19 grenade machine gun, Mod 3; improved M72 light antiarmor weapon; M136 AT4; M141 bunker defeat munition (BDM); and Javelin shoulder-fired munition.

M9 PISTOL

10-6. A lightweight, semiautomatic, single-action/double-action pistol can be unloaded without activating the trigger while the safety is in the "on" position (Figure 10-1). The M9 has a 15-round staggered magazine. The reversible magazine release button that can be positioned for either right-or left-handed shooters. This gun may be fired without a magazine inserted. The M9 is only authorized for 9-mm ball or dummy ammunition that is manufactured to US and NATO specifications. On this weapon, the hammer may be lowered from the cocked, "ready-to-fire" position to the uncocked position without activating the trigger. This is done by placing the thumb safety "ON."

Combat Marksmanship

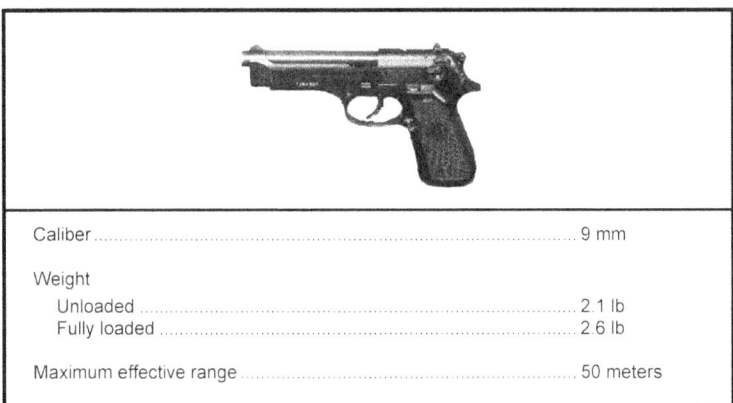

Caliber	9 mm
Weight	
Unloaded	2.1 lb
Fully loaded	2.6 lb
Maximum effective range	50 meters

Figure 10-1. M9 pistol.

M16-SERIES RIFLES

10-7. A lightweight, air-cooled, gas-operated, magazine-fed rifle designed for either burst or semiautomatic fire through use of a selector lever. There are three models:

M16A2

10-8. The M16A2 incorporates improvements in iron sight, pistol grip, stock, and overall combat effectiveness (Figure 10-2). Accuracy is enhanced by an improved muzzle compensator, a three-round burst control, and a heavier barrel; and by using the heavier NATO-standard ammunition, which is also fired by the squad automatic weapon (SAW).

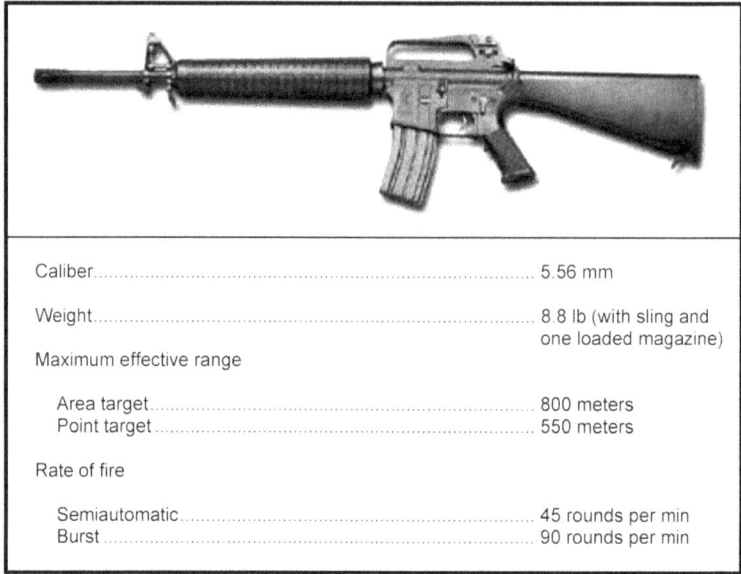

Caliber	5.56 mm
Weight	8.8 lb (with sling and one loaded magazine)
Maximum effective range	
Area target	800 meters
Point target	550 meters
Rate of fire	
Semiautomatic	45 rounds per min
Burst	90 rounds per min

Figure 10-2. M16A2 rifle.

M16A3

10-9. The M16A3 is identical to the M16A2, except the A3 has a removable carrying handle mounted on a picatinny rail (for better mounting of optics).

M16A4

10-10. The M16A4 is identical to the M16A3, except for the removable carrying handle mounted on a picatinny rail. It has a maximum effective range of 600 meters for area targets. Like the M4-series weapons, the M16-series rifles use ball, tracer, dummy, blank, and short-range training ammunition (SRTA) manufactured to US and NATO specifications.

M4 CARBINE

10-11. The M4 is a compact version of the M16A2 rifle, with collapsible stock, flat-top upper receiver accessory rail, and detachable handle/rear aperture site assembly (Figure 10-3). This rifle enables a Soldier operating in close quarters to engage targets at extended ranges with accurate, lethal fire. It achieves more than 85 percent commonality with the M16A2 rifle.

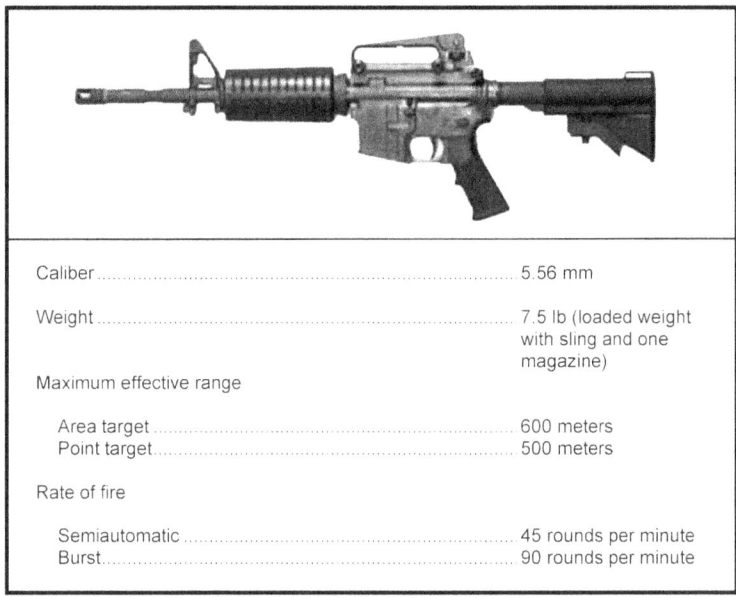

Caliber	5.56 mm
Weight	7.5 lb (loaded weight with sling and one magazine)
Maximum effective range	
Area target	600 meters
Point target	500 meters
Rate of fire	
Semiautomatic	45 rounds per minute
Burst	90 rounds per minute

Figure 10-3. M4 carbine.

M203 GRENADE LAUNCHER

10-12. The M203A1 grenade launcher is a single-shot weapon designed for use with the M4 series carbine. It fires a 40-mm grenade (Figure 10-4). Both have a leaf sight and quadrant sight. The M203 fires high-explosive (HE), high-explosive dual-purpose (HEDP) round, buckshot, illumination, signal, CS (riot control), and training practice (TP) ammunition. Two M203s are issued per Infantry squad.

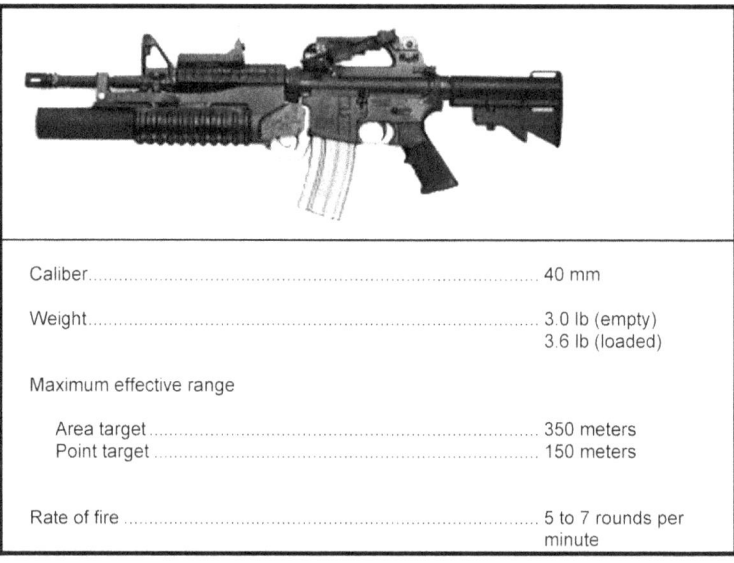

Caliber	40 mm
Weight	3.0 lb (empty) 3.6 lb (loaded)
Maximum effective range	
Area target	350 meters
Point target	150 meters
Rate of fire	5 to 7 rounds per minute

Figure 10-4. M203 grenade launcher.

M249 SQUAD AUTOMATIC WEAPON

10-13. A lightweight, gas-operated, air-cooled belt or magazine-fed, one-man-portable automatic weapon that fires from the open-bolt position (Figure 10-5). This gun can be fired from the shoulder, hip, or underarm position; from the bipod-steadied position; or from the tripod-mounted position. Two M249s are issued per Infantry squad.

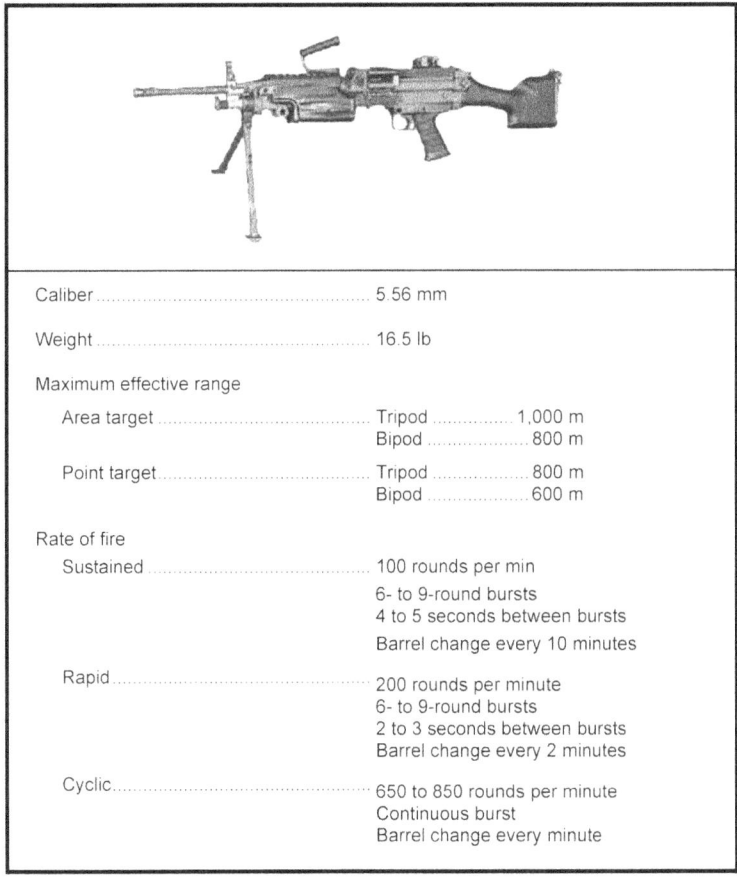

Caliber	5.56 mm
Weight	16.5 lb
Maximum effective range	
Area target	Tripod .... 1,000 m
	Bipod .... 800 m
Point target	Tripod .... 800 m
	Bipod .... 600 m
Rate of fire	
Sustained	100 rounds per min
	6- to 9-round bursts
	4 to 5 seconds between bursts
	Barrel change every 10 minutes
Rapid	200 rounds per minute
	6- to 9-round bursts
	2 to 3 seconds between bursts
	Barrel change every 2 minutes
Cyclic	650 to 850 rounds per minute
	Continuous burst
	Barrel change every minute

Figure 10-5. M249 squad automatic weapon (SAW).

M240B MACHINE GUN

10-14. A medium, belt-fed, air-cooled, gas-operated, crew-served, fully automatic weapon that fires from the open bolt position (Figure 10-6). Ammunition is fed into the weapon from a 100-round bandoleer containing ball and tracer (4:1 mix) ammunition with disintegrating metallic split-link belt. Other types of ammunition available include armor-piercing, blank, and dummy rounds. It can be mounted on a bipod, tripod, aircraft, or vehicle. A spare barrel is issued with each M240B, and barrels can be changed quickly as the weapon has a fixed head space. It is being issued to Infantry, armor, combat engineer, special force/rangers, and selected field artillery units that require medium support fires.

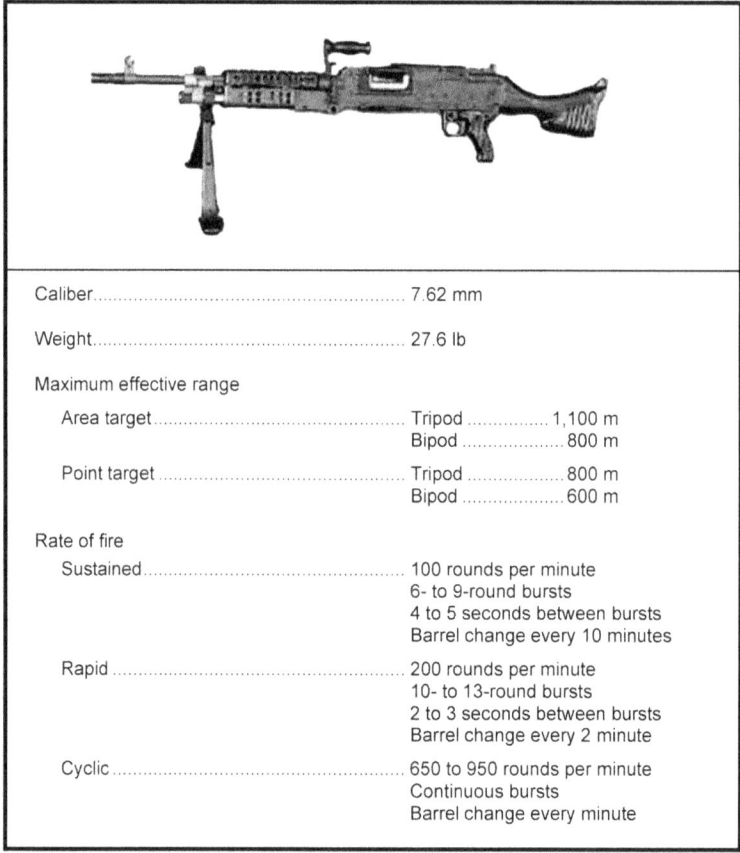

Caliber	7.62 mm
Weight	27.6 lb
Maximum effective range	
Area target	Tripod .... 1,100 m
	Bipod .... 800 m
Point target	Tripod .... 800 m
	Bipod .... 600 m
Rate of fire	
Sustained	100 rounds per minute
	6- to 9-round bursts
	4 to 5 seconds between bursts
	Barrel change every 10 minutes
Rapid	200 rounds per minute
	10- to 13-round bursts
	2 to 3 seconds between bursts
	Barrel change every 2 minute
Cyclic	650 to 950 rounds per minute
	Continuous bursts
	Barrel change every minute

Figure 10-6. M240B machine gun.

M2 (.50 CALIBER) MACHINE GUN

10-15. A heavy (barrel), recoil operated, air-cooled, crew-served, and transportable fully automatic weapon with adjustable headspace (Figure 10-7). A disintegrating metallic link belt is used to feed the ammunition into the weapon. This gun may be mounted on ground mounts and most vehicles as an antipersonnel and antiaircraft/light armor weapon. The gun is equipped with leaf-type rear sight, flash suppressor and a spare barrel assembly. Associated components are the M63 antiaircraft mount and the M3 tripod mount.

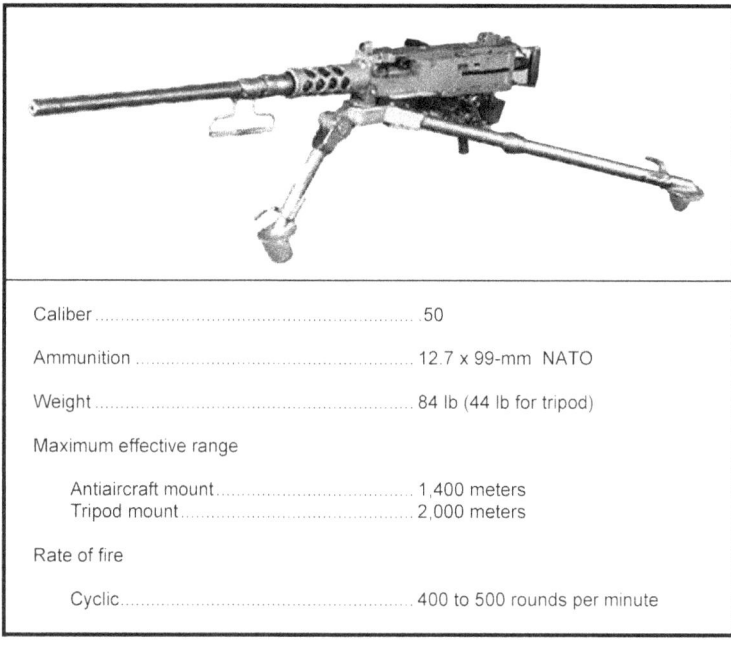

Caliber	.50
Ammunition	12.7 x 99-mm NATO
Weight	84 lb (44 lb for tripod)
Maximum effective range	
Antiaircraft mount	1,400 meters
Tripod mount	2,000 meters
Rate of fire	
Cyclic	400 to 500 rounds per minute

Figure 10-7. M2 .50 caliber machine gun with M3 tripod mount.

Chapter 10

MK 19 GRENADE MACHINE GUN, MOD 3

10-16. A self-powered, air-cooled, belt-fed, blowback-operated weapon designed to deliver decisive firepower against enemy personnel and lightly armored vehicles (Figure 10-8). A disintegrating metallic link belt feeds either HE or HEDP ammunition through the left side of the weapon. It replaces the M2 heavy machine guns in selected units, and will be the main suppressive weapon for combat support and combat service support units. The MK 19 Mod 3 can be mounted on the HMMWV M113 family of vehicles, on 5-ton trucks, and on some M88A1 recovery vehicles.

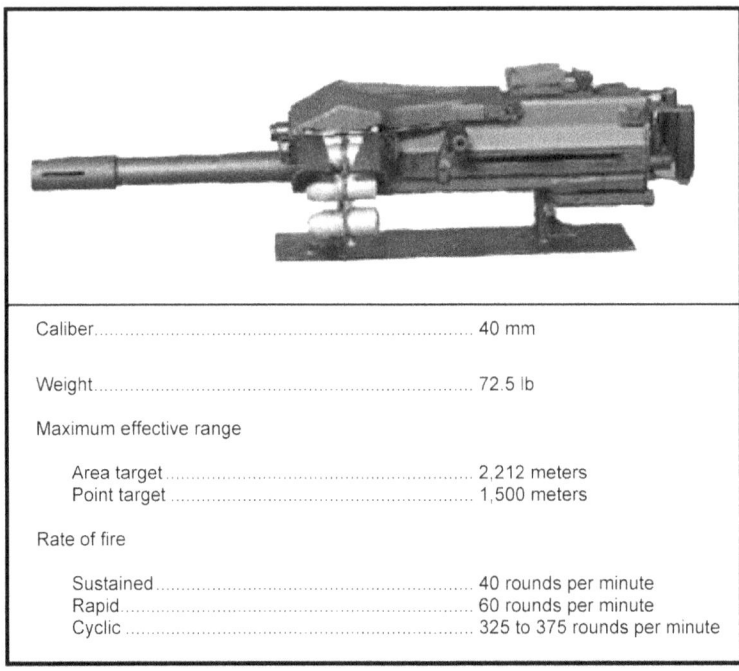

Caliber .... 40 mm

Weight .... 72.5 lb

Maximum effective range

    Area target .... 2,212 meters
    Point target .... 1,500 meters

Rate of fire

    Sustained .... 40 rounds per minute
    Rapid .... 60 rounds per minute
    Cyclic .... 325 to 375 rounds per minute

Figure 10-8. MK 19 grenade machine gun, Mod 3.

Combat Marksmanship

IMPROVED M72 LIGHT ANTIARMOR WEAPON

10-17. A compact, lightweight, single shot, and disposable weapon with a family of warheads optimized to defeat lightly armored vehicles and other hard targets at close-combat ranges (Figure 10-9). Issued as a round of ammunition, it requires no maintenance. The improved M72 light antiarmor weapon systems offer significantly enhanced capability beyond that of the combat-proven M72A3. The improved M72 light antiarmor weapon system consists of an unguided high-explosive antiarmor rocket prepackaged in a telescoping, throw-away launcher. The system performance improvements include a higher velocity rocket motor that extends the weapon effective range, increased lethality warheads, lower more consistent trigger release force, rifle type sight system, and better overall system reliability and safety. The improved M72 is transportable by tactical wheeled and tracked vehicles, without any safety constraints, and is air deliverable by individual parachutist or by pallet.

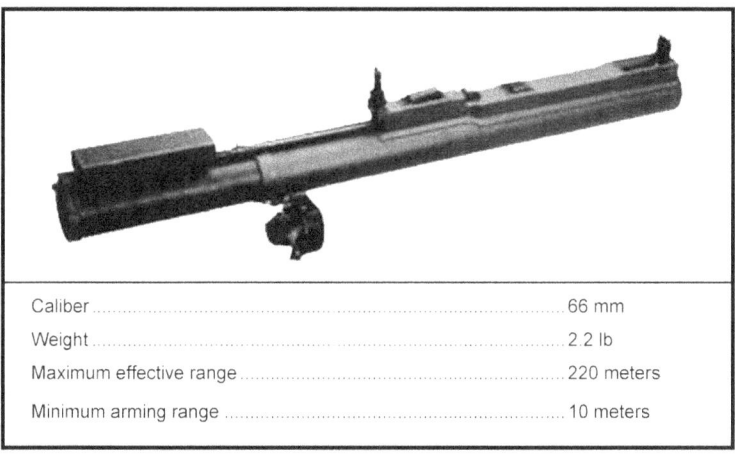

Caliber	66 mm
Weight	2.2 lb
Maximum effective range	220 meters
Minimum arming range	10 meters

Figure 10-9. Improved M72 LAW.

13 August 2013　　　TC 3-21.75　　　10-11

M136 AT4

10-18. The M136 AT4 is a lightweight, self-contained antiarmor weapon (Figure 10-10). It is man-portable and fires (only) from the right shoulder. The M136 AT4 is used primarily by Infantry forces to engage and defeat armor threats. The weapon accurately delivers a high-explosive antitank (HEAT) warhead with excellent penetration capability (more than 15 inches of homogenous armor) and lethal after-armor effects. The weapon has a free-flight, fin-stabilized, rocket-type cartridge packed in an expendable, one-piece, fiberglass-wrapped tube.

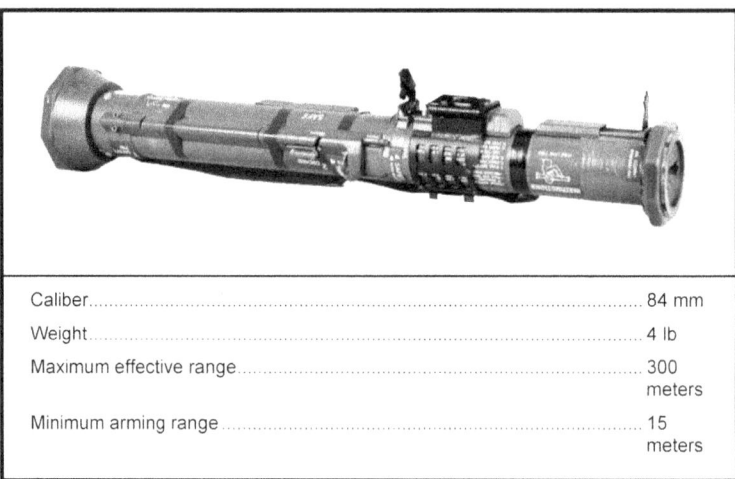

Caliber	84 mm
Weight	4 lb
Maximum effective range	300 meters
Minimum arming range	15 meters

Figure 10-10. M136 AT4.

Note: Both the high penetration (HP) and reduced sensitivity (RS) versions of the AT4 Confined Space (CS) weapon offer improved safety. Unlike the original AT4, these models can be fired safely from within a room or other protected enclosure. They also have much less backblast and launch signature. Both the AT4 CS-HP and CS-RS consist of shock resistant, fiberglass-reinforced launching tubes fitted with firing mechanisms, popup sights, carrying slings, protective covers, and bumpers.

Combat Marksmanship

M141 BUNKER DEFEAT MUNITION

10-19. A lightweight, self-contained, man-portable, high-explosive, disposable, shoulder-launched, multipurpose assault weapon-disposable (SMAW-D) that contains all gunner features and controls necessary to aim, fire, and engage targets (Figure 10-11). It can defeat fortified positions (bunkers) made of earth and timber; urban structures; and lightly armored vehicles. The M141 BDM is issued as an 83-mm, high-explosive, dual-mode, assault rocket round. It requires no maintenance. It fires (only) from the right shoulder. The weapon system structure consists of inner and outer filament-woven composite tubes, stored one inside the other to reduce carry length. Unlike the M136 AT4 launcher, the M141 BDM must be extended before firing.

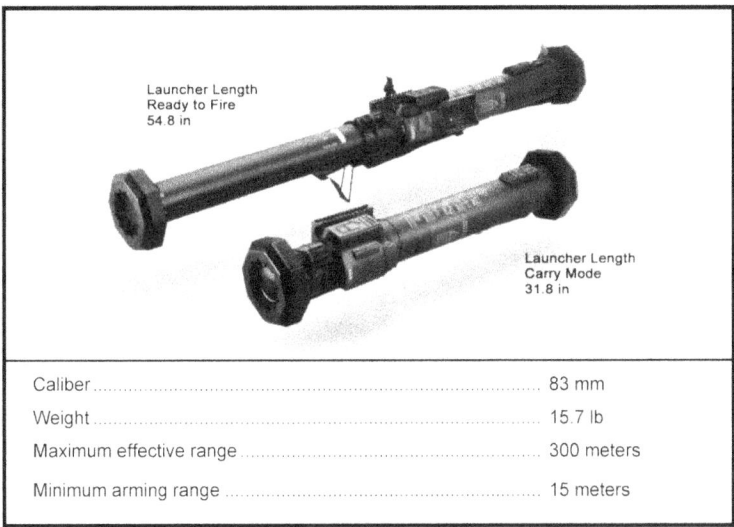

Caliber	83 mm
Weight	15.7 lb
Maximum effective range	300 meters
Minimum arming range	15 meters

Figure 10-11. M141 BDM.

Chapter 10

JAVELIN

10-20. The Javelin is the first *fire-and-forget*, crew-served antitank missile (Figure 10-12). Its fire-and-forget guidance mode enables gunners to fire and then immediately take cover. This greatly increases survivability. The Javelin's two major components are a reusable command launch unit (CLU) and a missile sealed in a disposable launch tube assembly. The CLU's integrated daysight/nightsight allows target engagements in adverse weather and in countermeasure environments. The CLU may also be used by itself for battlefield surveillance and reconnaissance. Special features include a selectable top-attack or direct-fire mode (for targets under cover or for use in urban terrain against bunkers and buildings), target lock-on before launch, and a very limited backblast. These features allow gunners to fire safely from within enclosures and covered fighting positions. The Javelin can also be installed on tracked, wheeled, or amphibious vehicles.

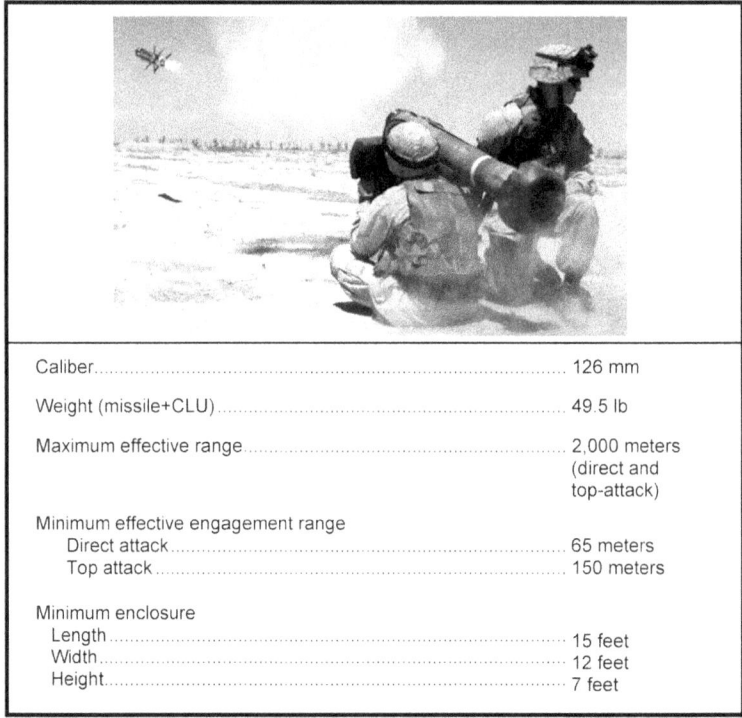

Caliber	126 mm
Weight (missile+CLU)	49.5 lb
Maximum effective range	2,000 meters (direct and top-attack)
Minimum effective engagement range	
Direct attack	65 meters
Top attack	150 meters
Minimum enclosure	
Length	15 feet
Width	12 feet
Height	7 feet

Figure 10-12. Javelin.

FIRE CONTROL

10-21. Fire control includes all actions in planning, preparing, and applying fire on a target. Your leader selects and designates targets. He also designates the midpoint and flanks or ends of a target, unless they are obvious for you to identify. When firing, you should continue to fire until the target is neutralized or until signaled to do otherwise by your leader. The noise and confusion of battle will limit the use of some methods of control, so he will use the method, or combination of methods, that does the job.

Combat Marksmanship

WAYS TO COMMUNICATE FIRE CONTROL

10-22. The following paragraphs discuss methods to correspond fire control.

Sound Signals

10-23. This includes both voice and devices such as whistles. Sound signals are good only for short distances. Their range and reliability are reduced by battle noise, weather, terrain, and vegetation. Voice communications may come directly from your leader to you or they may be passed from Soldier-to-Soldier.

Trigger Points/Lines

10-24. This method is prearranged fire; your leader tells you to start firing once the enemy reaches a certain point or terrain feature. When using this method of fire control, you do not have to wait for an order to start firing. Prearranged fire can also be cued to friendly actions.

Visual Signals

10-25. In this method, your leader gives a prearranged signal when he wants you to begin, shift, and cease firing. This can be either a visual signal or a sound signal. Start firing immediately when you get the signal.

Time

10-26. You may be instructed to begin, shift, and cease firing at a set time. Additionally, Soldier-initiated fire is used when there is no time to wait for orders from your leader.

Standing Operating Procedure

10-27. Using an SOP can reduce the number of oral orders needed to control fire. However, everyone in the unit must know and understand the SOP for it to work. Three widely used SOP formats are the search-fire-check, return-fire, and rate-of-fire SOPs. Procedures for issuing fire commands for direct fire weapons should also be covered in an SOP.

Search-Fire-Check SOP

1. Search your assigned sectors for enemy targets.
2. Fire at any targets (appropriate for your weapon) seen in your sectors.
3. While firing in your sectors, visually check with your leader for specific orders.

Return-Fire SOP

10-28. This SOP tells each Soldier in a unit what to do in case the unit makes unexpected contact with the enemy (in an ambush, for example).

Rate-Of-Fire SOP

10-29. This SOP tells each Soldier how fast to fire at the enemy. The rate of fire varies among weapons, but the principle is to fire at a maximum rate when first engaging a target and then slow the rate to a point that will keep the target suppressed. This helps to keep weapons from running out of ammunition.

Chapter 10

THREAT-BASED FIRE CONTROL MEASURES

10-30. The following paragraphs discuss threat-based fire control measures:

Engagement Priorities

10-31. Engagement priorities are the target types, identified by your leader, that offer the greatest payoff or present the greatest threat. He then establishes these as a unit engagement priority. Your leader refines these priorities within your unit, such as employing the best weapons for targets, as well as fire distribution.

Range Selection

10-32. Range selection is a means by which your leader will use their estimate of the situation to specify the range and ammunition for the engagement. Range selection is dependent on the anticipated engagement range. Terrain, visibility, weather, and light conditions affect range selection, and the amount and type of ammunition.

Weapons Control Status

10-33. The three levels of weapons control status outline the conditions, based on target identification criteria, under which friendly elements may engage. Your leader will set and adjust the weapons control status based on the friendly and enemy situation, and the clarity of the situation. The three levels, in descending order of restriction, follow:

WEAPONS HOLD   Engage only if engaged or ordered to engage.
WEAPONS TIGHT   Only engage targets that are positively identified as enemy.
WEAPONS FREE   Engage any targets that are not positively identified as friendly.

Rules of Engagement

10-34. ROE are the commander's rules for use of force and specify the circumstances and limitations in which you may use your weapon. They include definitions of combatant and noncombatant elements, and prescribe the treatment of noncombatants. Factors influencing ROE are national command policy, the mission, operational environment, and the law of war.

COMBAT ZERO

10-35. Combat readiness makes it essential for you to zero your individual weapon whenever it is issued. Additionally, each rifle in the unit arms room, even if unassigned, should be zeroed by the last Soldier it was assigned to. For more detailed information see FM 3-22.9.

Note: Only the M16/M4-series weapons will be discussed for zeroing procedures and aided-vision device combinations. For other weapons such as M249 SAW and the M240B, see the appropriate FMs.

MECHANICAL ZERO

10-36. Mechanical zero is the process of alignment of the weapons sighting systems to a common start point. Conduct the following procedures for these specific weapons:

Combat Marksmanship

M16A2/A3

10-37. Numbers in parentheses refer to the callouts in Figure 10-13.

1. Adjust the front sight post (1) up or down until the base of the front sight post is flush with the front sight post housing (2).
2. Adjust the elevation knob (3) counterclockwise, as viewed from above, until the rear sight assembly (4) rests flush with the carrying handle and the 8/3 marking is aligned with the index line on the left side of the carrying handle.
3. Position the apertures (5) so the unmarked aperture is up and the 0 to 200 meter aperture is down. Rotate the windage knob (6) to align the index mark on the 0 to 200 meter aperture with the long center index line on the rear sight assembly.

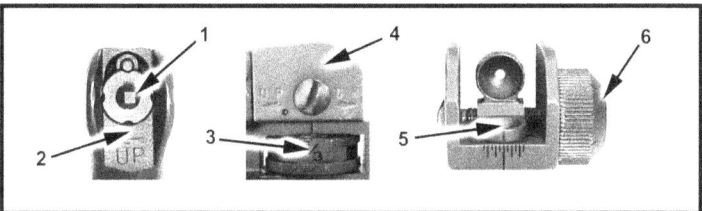

Figure 10-13. M16A2/A3 rifle mechanical zero.

M16A4 AND M4 CARBINE

10-38. Numbers in parentheses refer to the callouts in Figure 10-14.

1. Adjust the front sight post (1) up or down until the base of the front sight post is flush with the front sight post housing (2).
2. Adjust the elevation knob (3) counterclockwise, when viewed from above, until the rear sight assembly (4) rests flush with the detachable carrying handle and the 6/3 marking is aligned with the index line (5) on the left side of the carrying handle.
3. Position the apertures (6) so the unmarked aperture is up and the 0 to 200 meter aperture is down. Rotate the windage knob (7) to align the index mark (8) on the 0 to 200 meter aperture with the long center index line on the rear sight assembly.

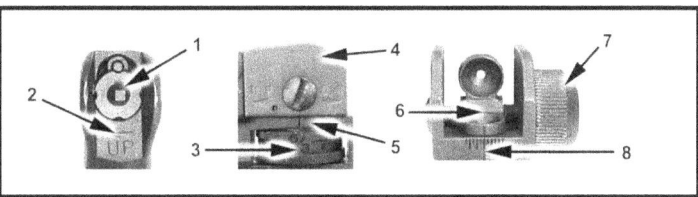

Figure 10-14. M16A4 and M4 carbine rifle mechanical zero.

BATTLESIGHT ZERO

10-39. Battlesight zero is the alignment of the sights with the weapon's barrel given standard issue ammunition. It provides the highest probability of hitting most high-priority combat targets with minimum

Chapter 10

adjustment to the aiming point 300 meter sight setting as on the M16A2/3/4 and M4 series weapons. For each of the following weapons, ensure the rear sights are set for battlesight zero (25-meter zero):

M16A2/A3

10-40. Numbers in parentheses refer to the callouts in Figure 10-15.

1. Adjust the elevation knob (1) counterclockwise, as viewed from above, until the rear sight assembly (2) rests flush with the carrying handle and the 8/3 marking is aligned with the index line (3) on the left side of the carrying handle. Then, turn the elevation knob one more click clockwise.
2. Position the apertures (4) so the unmarked aperture is up and the 0 to 200 meter aperture is down. Rotate the windage knob (5) to align the index mark on the 0 to 200 meter aperture with the long center index line on the rear sight assembly.

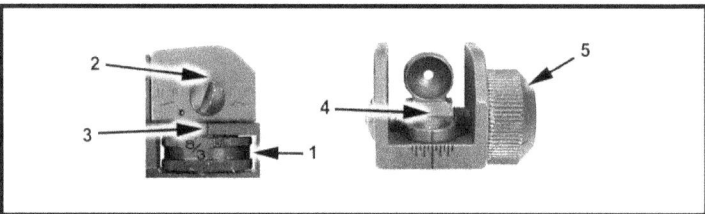

Figure 10-15. M16A2/A3 rifle battlesight zero.

M16A4

10-41. Numbers in parentheses refer to the callouts in Figure 10-16.

1. Adjust the elevation knob (1) counterclockwise, when viewed from above, until the rear sight assembly (2) rests flush with the detachable carrying handle and the 6/3 marking is aligned with the index line (3) on the left side of the detachable carrying handle. To finish the procedure, adjust the elevation knob two clicks clockwise so the index line on the left side of the detachable carrying handle is aligned with the "Z" on the elevation knob.
2. Position the apertures (4) so the unmarked aperture is up and the 0 to 200 meter aperture is down. Rotate the windage knob (5) to align the index mark on the 0 to 200 meter aperture with the long center index line (6) on the rear sight assembly.

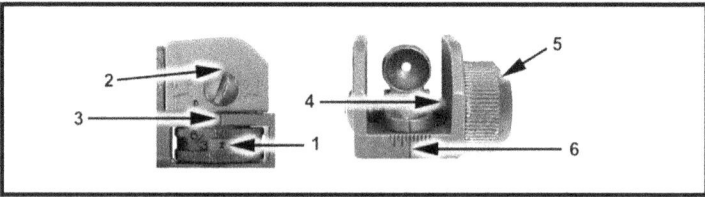

Figure 10-16. M16A4 rifle battlesight zero.

10-18　　　　　　　　　　　　TC 3-21.75　　　　　　　　　　　　13 August 2013

Combat Marksmanship

M4

10-42. Numbers in parentheses refer to the callouts in Figure 10-17.

1. Adjust the elevation knob (1) counterclockwise, when viewed from above, until the rear sight assembly (2) rests flush with the detachable carrying handle and the 6/3 marking is aligned with the index line (3) on the left side of the detachable carrying handle. The elevation knob remains flush.
2. Position the apertures (4) so the unmarked aperture is up and the 0 to 200 meter aperture is down. Rotate the windage knob (5) to align the index mark (6) on the 0 to 200 meter aperture with the long center index line on the rear sight assembly.

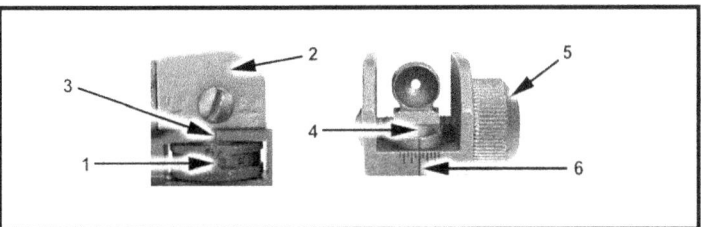

Figure 10-17. M4 rifle battlesight zero.

SHOT GROUPS

10-43. To ensure proper and accurate shot group marking--

1. Apply the four fundamentals of marksmanship deliberately and consistently. Establish a steady position allowing observation of the target. Aim the rifle at the target by aligning the sight system, and fire your rifle without disturbing this alignment by improper breathing or during trigger squeeze.
2. Initially, you should fire two individual shot groups before you consider changing the sight. Fire each shot at the same aiming point (center mass of the target) from a supported firing position. You will fire a three-round shot group at the 25-meter zero target.
3. You will triangulate each shot group and put the number "1" in the center of the first shot group and a number "2" on the second. Group the two shot groups and mark the center of the two shot groups with an X. If the two shot groups fall within a 4-centimeter circle then determine what sight adjustments need to be made, then identify the closest horizontal and vertical lines to the X, and then read the 25-meter zero target to determine the proper sight adjustments to make. A proper zero is achieved if five out of six rounds fall within the 4-centimeter circle (Figure 10-18).

Chapter 10

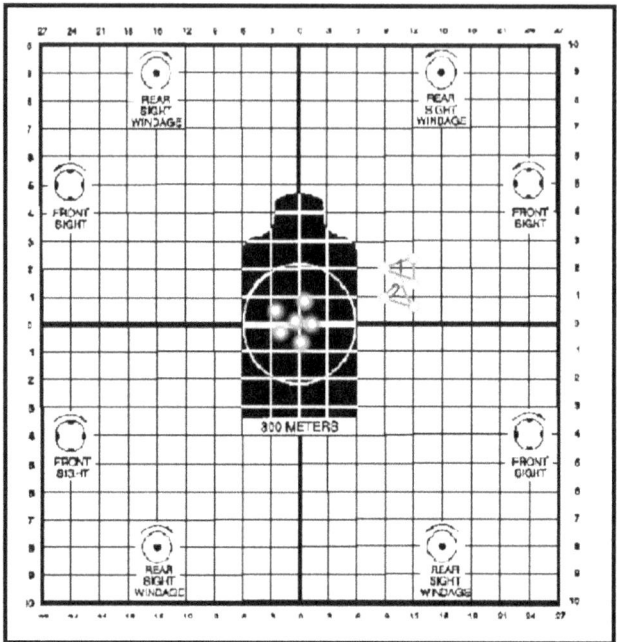

Figure 10-18. Final shot group results.

Note: The M16A2 zero target squares are .96 centimeter in size while the M4 zero target squares are 1.3 centimeters in size. Two single shots on a 25-meter zero target that are 2 centimeters apart does not equate to two squares from each other on the M4 zero target.

Combat Marksmanship

BORELIGHT ZERO

10-44. The borelight is an accurate means of zeroing weapons and most aided-vision equipment without the use of ammunition. Time and effort must be applied to ensure a precise boresight, which will in turn save time and ammunition. Both the M16A2 and the M4/MWS can be zeroed using the borelight and each of the following five aided vision devices:

- Backup iron sight (which can also be boresighted).
- AN/PAQ-4B/C.
- AN/PEQ-2A.
- AN/PAS-13.
- M68 CCO.

Note: Precisely boresighting a laser allows direct engagement of targets without a 25-meter zero. If a borelight is unavailable, you must use a 25-meter zero to zero the device. All optics must be 25-meter zeroed. A borelight only aids in zeroing.

DANGER

1. DO NOT STARE INTO THE VISIBLE LASER BEAM.
2. DO NOT LOOK INTO THE VISIBLE LASER BEAM THROUGH BINOCULARS OR TELESCOPES.
3. DO NOT POINT THE VISIBLE LASER BEAM AT BEAM THROUGH BINOCULARS OR TELESCOPES.
4. DO NOT SHINE THE VISIBLE LASER BEAM INTO OTHER INDIVIDUAL'S EYES.

WARNINGS

1. Make sure the weapon is CLEAR and on SAFE before using the borelight.
2. Ensure the bolt is locked in the forward position.
3. When rotating the borelight to zero it, ensure the mandrel is turning counter clockwise (from the gunners point of view) to avoid loosening the borelight from the mandrel.

10-45. Boresighting is a simple procedure if the steps are strictly followed. The visible laser of the borelight must be aligned with the barrel of your assigned weapon. Then, using a 10-meter boresight target, the weapon can be boresighted with any optic, laser, or iron sight.

10-46. Before you boresight your weapon, the borelight must first be zeroed to the weapon. To zero the borelight to the weapon, align the visible laser with the barrel of the weapon. Stabilizing the weapon is crucial. The weapon can be stabilized in a rifle box rest or in a field location by laying two rucksacks side by side. Lay the weapon on the rucksacks and then lay another rucksack on top of the weapon to stabilize it.

Chapter 10

ZEROING THE BORELIGHT

> **WARNING**
>
> Avoid overadjusting the laser or pointing it at other Soldiers or reflective material.

10-47. The weapon need not be perfectly level with the ground during boresighting. Conduct the following steps to zero the borelight:

1. Attach the 5.56-mm mandrel to the borelight.
2. Insert the mandrel into the muzzle of the weapon. The borelight is seated properly when the mandrel cannot be moved any further into the muzzle and the mandrel spins freely. Stabilize the weapon so it will not move.
3. Measure 10 meters with the 10-meter cord that comes with the borelight or pace off eleven paces.
4. The zeroing mark is a small dot drawn on a piece of paper, tree bark, or the borelight RP on the 10-meter boresight target (Figure 10-19).

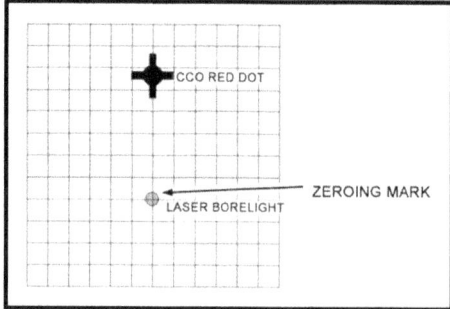

Figure 10-19. Example zeroing mark.

5. Rotate the borelight until the battery compartment is facing upward and the adjusters are on the bottom (Figure 10-20). This position of the borelight, and where the visible laser is pointing, is the start point.

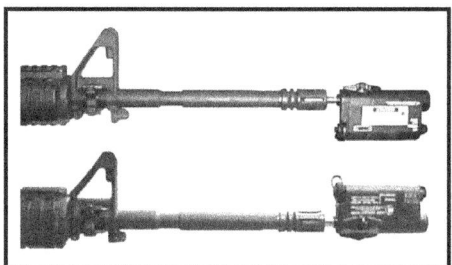

Figure 10-20. Borelight in the start point position.

6. Rotate the borelight until the battery compartment is down and the adjusters are on top to allow for easy access to the adjusters (Figure 10-21). This position of the borelight, and where the visible laser is pointing, is identified as the half-turn position.

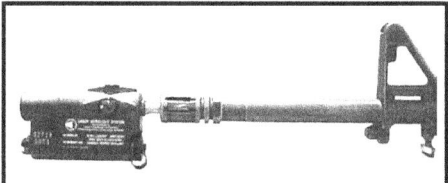

Figure 10-21. Borelight in the half-turn position.

Note: Use the commands *START POINT* and *HALF TURN* to ensure clear communication between you and your buddy/assistant at the boresight target.

Chapter 10

7. The RP is about halfway between the start point and the half-turn point (Figure 10-22).

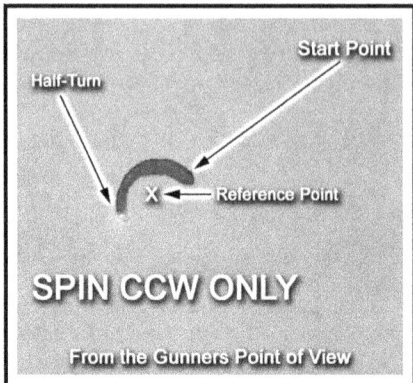

Figure 10-22. Examples of start point, half-turn, and reference point.

8. Turn the borelight on and spin it until it is in the start point position. Place the zeroing mark about 10 meters from the end of the barrel so that the visible laser strikes the zeroing mark.
9. Slowly rotate the borelight 180 degrees while watching the visible laser made by the borelight. If the visible laser stops on the zeroing mark, the borelight is zeroed to the weapon.
10. If the borelight does not stop on the zeroing mark, elevation and windage adjustments must be made to the borelight.
11. From the start point, realign the zeroing mark with the visible laser, rotate the borelight 180 degrees to the half-turn position, and identify the RP. Using the adjusters on the borelight, move the visible laser to the RP. Rotate the borelight back to the start point; move the zeroing mark to the visible laser. If the visible laser cannot be located when you spin the borelight to the half-turn position, start this procedure at 2 meters instead of 10 meters. When the visible laser is adjusted to the RP at 2 meters, then start the procedure again at 10 meters.
12. Repeat Step 11 above until the visible laser spins on itself.

Note: Every barrel is different; therefore, steps (8) through (10) must be performed with every weapon to ensure that the borelight is zeroed to that barrel. If the borelight is zeroed, then go directly to the boresighting procedures.

Combat Marksmanship

10-48. During boresighting, your weapon should be in the "bolt forward" position and must not be canted left or right. As a firer, you will need a target holder in order to properly boresight a weapon. The duties follow:

1. The firer's primary duty is to zero the borelight and make all adjustments on the aided-vision device being used.
2. The target holder secures the 10-meter, (1-centimeter square grid) boresight target (Figure 10-23) straight up and down 10 meters from the borelight, and directs the firer in making necessary adjustments to the aiming device. The target holder must wear NVGs when boresighting IR aiming lasers.

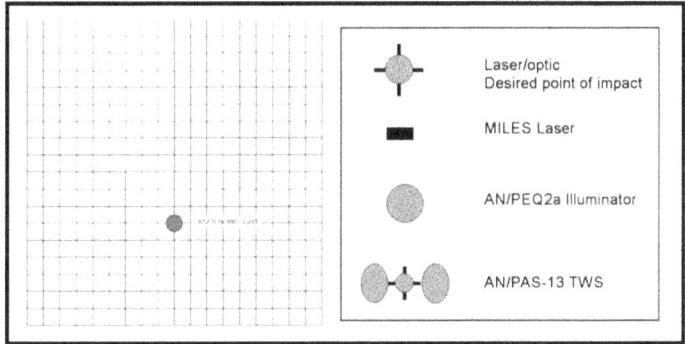

Figure 10-23. Blank 10-meter boresight target and offset symbols.

BORESIGHTING THE BACKUP IRON SIGHT

10-49. The backup iron sight (BIS) can be boresighted to a new user to expedite 25-meter zeroing (Figure 10-24). To boresight using the BIS, align the iron sights with the (Canadian bull) on the 10-meter boresight target. Adjust the windage and elevation of the iron sights until the borelight is centered with the circle on the boresight target.

10-50. The BIS is adjusted for a 300-meter battlefield zero to provide backup in the event an optic or laser device fails to function. The BIS is zeroed on the M4/M4A1 target on the backside of the M16A2 zero targets. The 25-meter zeroing procedures are the same as for conventional rear sight assembly on the M16-/M4-series weapons.

Figure 10-24. Backup iron sight.

Chapter 10

BORESIGHTING THE M68 CLOSE COMBAT OPTIC

10-51. The M68 CCO is a reflex (nontelescopic) sight (Figure 10-25). It uses a red-aiming reference (collimated dot) and is designed for the sighting with both eyes open. Position your head so that one eye can focus on the red dot and the other eye can scan downrange. Place the red dot on the center of mass of the target and engage. With your nonfiring eye closed, look through the M68 to ensure you can see the red dot clearly. Place the red dot center mass of the target, and then engage. If you zero your weapon with the one-eye-open method, then you must engage targets using this method for zero accuracy. The dot follows the horizontal and vertical movement of your eye while remaining fixed on the target. No centering or focusing is required. The more accurately you boresight the M68 to your weapon, the closer it will be to a battlesight zero.

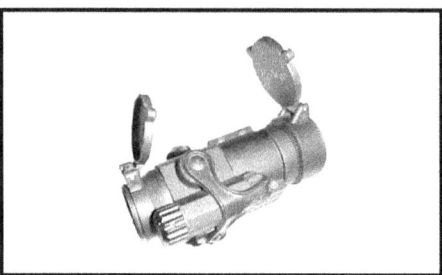

Figure 10-25. M68 close combat optic.

25-Meter Zero Procedure with Borelight

1. Select the proper 10-meter boresight target for your weapon and M68 configuration. With the help of an assistant, place the boresight target 10 meters in front of the weapon.

2. Turn the M68 to the desired setting (position number 4). Assume a stable supported firing position behind your weapon, and look through the M68. Aim the red dot of the M68 on the crosshair of the 10-meter boresight target. Adjust the M68 until the borelight's visible laser centers on the borelight circle on the 10-meter boresight target.

3. Turn the borelight off. Move your weapon off the crosshair, realign the red dot of the M68 on the crosshair, and turn the borelight back on. If the borelight is on the circle and the red dot of the M68 is on the crosshair, your weapon system is now boresighted.

Note: The M68 is parallax free beyond 50 meters. Boresight at 10 meters. In order to get a solid boresight, acquire the same sight picture and cheek-to-stock weld position each time.

4. Turn the laser off. To avoid damaging the borelight device, carefully remove it, and the mandrel, from the weapon.

25-Meter Zero Procedure without Borelight

10-52. If the borelight is unavailable, conduct a 25-meter zero--

1. Use the same standards as with iron sights.
2. Starting from center mass of the 300-meter silhouette on the 25-meter zero target, count down 1.4 centimeters and make a mark. This is now the point of impact.
3. Make a 4-by-4 square box around the point of impact. This box is now the offset and is the designated point of impact for the M68.
4. Aim center mass of the 300-meter silhouette and adjust to the M68 so that the rounds impact in the 4-by-4 square box, 1.4 centimeters down from the point of aim.
    - Two clicks = 1 cm at 25 m for windage and elevation.
    - One click clockwise on elevation moves bullet strike down.
    - One click clockwise on windage moves bullet strike left.
5. Zero only on the M16A2 25-meter zero target.

Note: At ranges of 50 meters and beyond, parallax is minimal. However, at ranges of 50 meters and closer, keep the red dot centered while zeroing. Use the same aiming method (one or both eyes open) for zeroing that you plan to use to engage targets.

WARNING

Check the light for proper intensity before opening the front lens cover. Close the front lens cover before turning the rotary switch counterclockwise to the OFF position. Failure to follow this warning could reveal your position to the enemy.

BORESIGHTING AN/PAS-13 (V2) AND (V3) THERMAL WEAPON SIGHTS

10-53. Boresight and zero both the narrow and wide FOVs. Zero at 25 meters to ensure you have zeroed the TWS properly:

1. Select the proper 10-meter boresight target for the weapon/TWS configuration and, with the help of an assistant, place the boresight target 10 meters in front of the weapon.
2. Ensure the M16/M4 reticle displays. Assume a stable supported firing position behind your weapon, and look through the TWS.
3. Place a finger on each oval of the 10-meter boresight target. Aim between your fingers with the 300-meter aiming point. Adjust the TWS until the borelight's visible laser centers on the borelight circle's 10-meter boresight target.
4. Have gunner move off the aiming block, realign the TWS to the center of the heated block, and then turn the borelight back on. Ensure you still have the proper boresight alignment. Now you are boresighted.
5. Change the FOV on the sight by rotating the FOV ring, and repeat steps (1) through (4).
6. Turn the laser off. To avoid damaging the borelight device, carefully remove it and the mandrel from the weapon.

25-Meter Zero Procedure

10-54. Ensure you zero both FOVs.

1. Use the same procedures and standards as with iron sights (Figure 10-26).
2. At the 25-meter range, each increment of azimuth or elevation setting moves the strike of the round as follows:
   -- 1 1/4 centimeters for medium TWS on wide FOV.
   -- 3/4 centimeter for MTWS on narrow FOV.
   -- 3/4 centimeter for heavy TWS on WFOV.
   -- 1/4 centimeter for HTWS on NFOV.
3. Retighten the rail grabber after you fire the first three rounds.

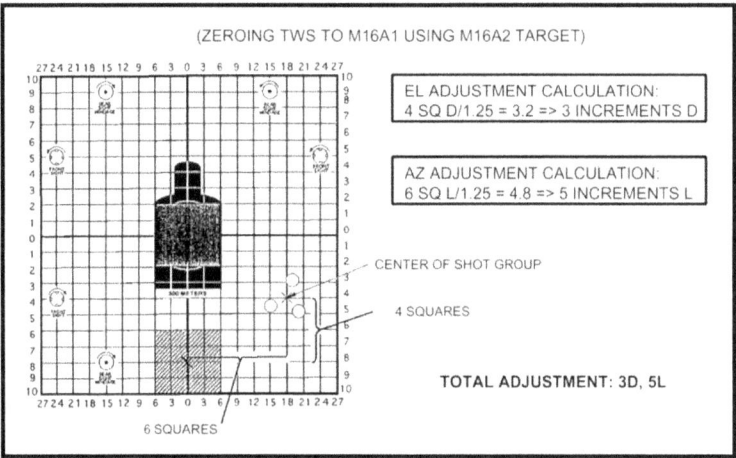

Figure 10-26. Example TWS zeroing adjustments.

Boresighting AN/PAQ-4B/C

1. Select the proper 10-meter boresight target for the weapon-to-AN/PAQ-4B/C configuration and, with the help of an assistant, place the boresight target 10 meters in front your weapon.

2. Install the borelight filter and turn on the AN/PAQ-4B/C. Align the 10-meter boresight target with the visible laser of the borelight.

3. Adjust the adjusters on the AN/PAQ-4B/C until the IR laser is centered on the crosshair located on the 10-meter boresight target.

   -- Keep the boresight target and zeroing mark stable during the boresight procedure.

   -- Do not turn the adjustment screws too much or they will break. Regardless of the mounting location, the adjuster that is on top or bottom will always be the adjuster for elevation and the one on the side will be the windage adjuster.

   -- Elevation adjustment screw—one click at 25 meters = 1 centimeter.

   -- Windage adjustment screw—one click at 25 meters = 1 centimeter.

Combat Marksmanship

25-METER ZERO PROCEDURES

10-55. If the borelight is unavailable, conduct a 25-meter zero (Figure 10-27) as follows:

1. Use the same standards as for the iron sights.

2. Preset the adjusters IAW TM 11-5855-301-12&P.

3. Prepare 25-meter zero target by cutting a 3x3-centimeter square out of the center of the silhouette.

   Elevation Adjustment Screw—one click at 25 m = 1 cm (clockwise = up).
   Windage Adjustment Screw—one click at 25 m = 1 cm (clockwise = left).

4. Retighten rail grabber after the first three rounds are fired.

Note: When cutting the 3-centimeter square out of the target, some of the strike zone may be cut out. Exercise care when annotating the impact of the rounds. When the weapon is close to being zeroed, some of the shots may be lost through the hole in the target.

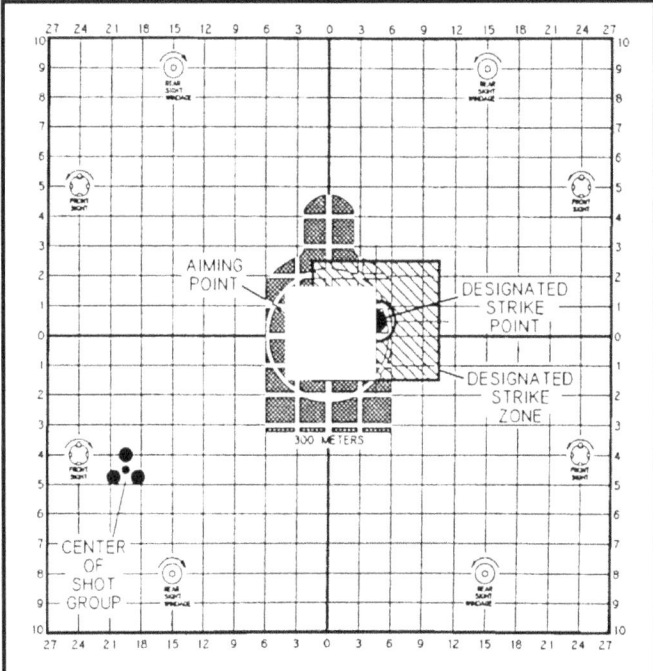

Figure 10-27. Example shot group adjustment with strike zone.

Chapter 10

BORESIGHTING AN/PEQ-2A

1. Select the proper 10-meter boresight target for the weapon and AN/PEQ-2A configuration and, with the help of an assistant, place the boresight target 10 meters in front your weapon.
2. Install the filter on the aiming laser and turn on the AN/PEQ-2A. Align the 10-meter boresight target with the visible laser of the borelight.
3. Adjust the adjusters on the AN/PEQ-2A until the IR laser centers on the crosshair located on the 10-meter boresight target.
4. Adjust the illuminator in the same manner.
5. Turn the laser off. To avoid damaging the borelight device, carefully remove the borelight and the mandrel from the weapon.
    -- Each click of elevation and windage is 1 centimeter. For ease, round up one square. However, each square of the 25-meter zero target is 0.9 centimeter, which affects large adjustments.
    -- Do not turn the adjustment screws too much, or they will break. Regardless of the mounting location, the adjuster that is on top or bottom will always be the adjuster for elevation and the one on the side will be the windage adjuster.

25-METER ZERO PROCEDURES

10-56. If a borelight is unavailable, you must conduct a 25-meter zero:

1. Follow the same standards as you did with iron sights.
2. Preset the adjusters IAW TM 11-5855-308-12&P.
3. Prepare the 25-meter zero target by cutting out a 3x3-centimeter square in the center of the target and E-type silhouette.
4. Turn the aiming beam on in the low power setting (AIM LO). Install aim point filter to eliminate excessive blooming. The adjustments for the AN/PEQ-2A (top mounted) follow:

Aiming Point
   Elevation Adjustment Screw--One click at 25 m = 1 cm or one square (clockwise = up).
   Windage Adjustment Screw--One click at 25 m = 1 cm or one square (clockwise = right).

Target Illuminator
   Elevation Adjustment Screw--One click at 25 m = 1 cm or one square (clockwise = down).
   Windage Adjustment Screw--One click at 25 m = 1 cm or one square (clockwise = right).

5. Retighten rail grabber and AN/PEQ-2A.
6. Once you have zeroed the aiming beam, rotate the selector knob to the DUAL LO, DUAL LO/HI, or DUAL HI/HI mode, and observe both aiming and illumination beams. Rotate the illumination beam adjusters to align the illumination beam with the aiming beam.

Note: To ensure zero retention--

1. Ensure you fully tighten the mounting brackets and the AN/PEQ-2A thumbscrew *prior* to zeroing.
2. Remove the TPIAL and rail grabber as a whole assembly, and then place it back onto the same notch.

MISFIRE PROCEDURES AND IMMEDIATE ACTION

MISFIRE

10-57. A misfire is the failure of a chambered round to fire. Ammunition defects and faulty firing mechanisms can cause misfires.

STOPPAGE

10-58. A stoppage is a failure of an automatic or semiautomatic firearm to complete the cycle of operation. You may apply immediate or remedial action to clear the stoppage. Some stoppages cannot be cleared by immediate or remedial action and may require weapon repair to correct the problem. Immediate action involves quickly applying a possible correction to reduce a stoppage without looking for the actual cause. Remedial action is the action taken to reduce a stoppage by looking for the cause and to try to clear the stoppage once it has been identified. To reduce a stoppage--

M9 Pistol

Immediate Action

10-59. Take immediate action is taken within 15 seconds of a stoppage.

1. Ensure the decocking/safety lever is in the FIRE position.
2. Squeeze the trigger again.
3. If the pistol does not fire, ensure the magazine is fully seated, retract the slide to the rear, and release.
4. Squeeze the trigger.
5. If the pistol does not fire again, remove the magazine and retract the slide to eject the chambered cartridge. Insert a new magazine, retract the slide, and release to chamber another cartridge.
6. Squeeze the trigger.
7. If the pistol still does not fire, perform remedial action.

Remedial Action

10-60. Remedial action is taken to reduce a stoppage by looking for the cause.

1. Clear the pistol.
2. Inspect the pistol for the cause of the stoppage.
3. Correct the cause of the stoppage, load the pistol, and fire.
4. If the pistol fails to fire again, disassemble it for closer inspection, cleaning, and lubrication.

Chapter 10

M16A2/3/4 And M4 Carbine Rifles

Immediate Action

10-61. Use the key word SPORTS to help you remember the steps to apply immediate action:

SPORTS

S LAP — gently upward on the magazine to ensure it is fully seated and the magazine follower is not jammed.

P ULL — the charging handle fully to the rear.

O BSERVE — for the ejection of a live round or expended cartridge. *

R ELEASE — the charging handle (do not ride it forward).

T AP — the forward assist assembly to ensure bolt closure.

S QUEEZE — the trigger and try to fire the rifle.

* If the weapon fails to eject a cartridge, perform remedial action.

Remedial Action

10-62. To apply the corrective steps for remedial action, first try to place the weapon on SAFE, then remove the magazine, lock the bolt to the rear, and place the weapon on safe.

M249 SAW and M240B Machine Guns

Immediate Action

10-63. If either weapon stops firing, the same misfire procedures will apply for both. You will use the keyword POPP, which will help you remember the steps in order. While keeping the weapon on your shoulder, Pull and lock the charging handle to the rear while Observing the ejection port to see if a cartridge case, belt link, or round is ejected. Ensure the bolt remains to the rear to prevent double feeding if a round or cartridge case is not ejected. If a cartridge case, belt link, or round is ejected, Push the charging handle to its forward position, take aim on the target, and Press the trigger. If the weapon does not fire, take remedial action. If a cartridge case, belt link, or round is not ejected, take remedial action.

Remedial Action

10-64. If immediate action does not remedy the problem, the following actions may be necessary to restore the weapon to operational condition:

Combat Marksmanship

Cold Weapon Procedures

10-65. When a stoppage occurs with a cold weapon, and if immediate action has failed--

1. While in the firing position, grasp the charging handle with your right hand, palm up; pull the charging handle to the rear, locking the bolt. While keeping resistance on the charging handle, move the safety to SAFE and return the cocking handle.
2. Place the weapon on the ground or away from your face. Open the feed cover and perform the five-point safety check. Reload and continue to fire.
3. If the weapon fails to fire, clear it, and inspect the weapon and the ammunition.

Hot Weapon Procedures

10-66. If the stoppage occurs with a hot weapon (200 or more rounds in less than 2 minutes, or as noted previously for training)--

1. Move the safety to SAFE and wait 5 seconds. During training, let the weapon cool for 15 minutes.
2. Use Cold Weapon Procedures 1 through 3 above.

Jammed Charging Handle

10-67. If a stoppage occurs and the charging handle cannot be pulled to the rear by hand (the bolt may be fully forward and locked or only partially forward) take the following steps:

1. Try once again to pull the charging handle by hand.
2. If the weapon is hot enough to cause a cook-off, move all Soldiers a safe distance from the weapon and keep them away for 15 minutes.
3. After the gun has cooled, open the cover and disassemble the gun. Ensure rearward pressure is kept on the charging handle until the buffer is removed. (The assistant gunner can help you do this.)
4. Remove the rounds or fired cartridges. Use a cleaning rod or ruptured cartridge extractor if necessary.

   -- In a training situation, after completing the remedial action procedures, do not fire the gun until an ordnance specialist has conducted an inspection.
   -- In a combat situation, after the stoppage has been corrected, you may change the barrel and try to fire. If the weapon fails to function properly, send it to the unit armorer.

REFLEXIVE FIRE

10-68. Reflexive fire is the automatic trained response to fire your weapon with minimal reaction time. Reflexive shooting allows little or no margin for error. Once you master these fundamentals, they will be your key to survival on the battlefield:

- Proper firing stance.
- Proper weapon ready position.
- Aiming technique.
- Aim point.
- Trigger manipulation.

Chapter 10

PROPER FIRING STANCE

10-69. Regardless of the ready position used, always assume the correct firing stance to ensure proper stability and accuracy when engaging targets. Keep your feet about shoulder-width apart. Toes are pointed straight to the front (direction of movement). The firing side foot is slightly staggered to the rear of the nonfiring side foot. Knees are slightly bent and the upper body is leaned slightly forward. Shoulders are square and pulled back, not rolled over or slouched. Keep your head up and both eyes open. When engaging targets, hold the weapon with the butt of the weapon firmly against your shoulder and the firing side elbow close against the body.

PROPER WEAPON READY POSITION

10-70. The two weapon ready positions are the high ready and low ready (Figure 10-28).

Low Ready Position--Place the butt of the weapon firmly in the pocket of your shoulder with the barrel pointed down at a 45-degree angle. With your nonfiring hand, grasp the handguards toward the front sling swivel, with your trigger finger outside the trigger well, and the thumb of your firing hand on the selector lever. To engage a target from this position, bring your weapon up until you achieve the proper sight picture. This technique is best for moving inside buildings.

High Ready Position--Hold the butt of the weapon under your armpit, with the barrel pointed slightly up so that the top of the front sight post is just below your line of sight, but within your peripheral vision. With your nonfiring hand, grasp the handguards toward the front sling swivel. Place your trigger finger outside the trigger well, and the thumb of your firing hand on the selector lever. To engage a target from this position, just push the weapon forward as if to bayonet the target and bring the butt stock firmly against your shoulder as it slides up your body. This technique is best suited for the lineup outside of a building, room, or bunker entrance.

Figure 10-28. Ready positions.

Combat Marksmanship

AIMING TECHNIQUES

10-71. The four aiming techniques with iron sights all have their place during combat in urban areas, but the *aimed quick-kill* technique is often used in precision room clearing. You need to clearly understand when, how, and where to use each technique.

Slow Aimed Fire--This technique is the slowest but most accurate. Take a steady position, properly align your sight picture, and squeeze off rounds. Use this technique only to engage targets beyond 25 meters when good cover and concealment is available, or when your need for accuracy overrides your need for speed.

Rapid Aimed Fire--This technique uses an imperfect sight picture. Focus on the target and raise your weapon until the front sight post assembly obscures the target. Elevation is less critical than windage when using this technique. This aiming technique is extremely effective on targets from 0 to 15 meters and at a rapid rate of fire.

Aimed Quick Kill--The aimed quick kill technique is the quickest and most accurate method of engaging targets up to 12 meters and greater. When using this technique, you must aim over the rear sight, down the length of the carry handle, and place the top 1/2 to 3/4 of an inch of the front sight post assembly on the target.

Instinctive Fire--This is the least accurate technique and should only be used in emergencies. It relies on your instinct, experience, and muscle memory. In order to use this technique, first concentrate on the target and point your weapon in the general direction of the target. While gripping the handguards with your nonfiring hand, extend your index finger to the front, automatically aiming the weapon on a line towards the target.

AIM POINT

10-72. Short-range engagements fall into two categories based on the mission and hostile threat.

Lethal Shot Placement--The lethal zone of the target is center mass between the waist and the chest. Shots in this area maximize the *hydrostatic shock* of the round.

Incapacitating Shot Placement --The only shot placement that guarantees immediate and total incapacitation is one roughly centered in the face. Shots to the side of the head should be centered between the crown of the skull and the middle of the ear opening, and from the center of the cheekbones to the middle of the back of the head.

TRIGGER MANIPULATION

10-73. Due to the reduced reaction time, imperfect sight picture, and requirement to effectively place rounds into threat targets, you must fire multiple rounds during each engagement in order to survive. Multiple shots may be fired using the controlled pair, automatic weapons fire, and the failure drill methods.

Controlled Pair--Fire two rounds rapid succession. When you fire the first, let the shot move the weapon in its natural arc and do not fight the recoil. Rapidly bring the weapon back on target and fire the second round. Fire controlled pairs at an individual target until he goes down. When you have multiple targets, fire a controlled pair at each target, and then reengage any targets left standing.

Automatic Fire--You might need automatic weapons fire to maximize violence of action or when you need fire superiority to gain a foothold in a room, building, or trench. You should be able to fire six rounds (two three-round bursts) in the same time it takes to fire a controlled pair. The accuracy of engaging targets can be equal to that of semiautomatic fire at 10 meters.

Failure Drill--To make sure a target is completely neutralized, you will need to be trained to execute the failure drill. Fire a controlled pair at the lethal zone of the target, and then fire a single shot to the incapacitating zone. This increases the probability of hitting the target with the first shot, and allows you to incapacitate him with the second shot.

This page intentionally left blank.

Chapter 11
Communications

Command and control is a vital function on the battlefield. Effective communications are essential to command and control. Information exchanged by two or more parties must be transmitted, received, and understood. Without it, units cannot maneuver effectively and leaders cannot command and control their units, which may result in lives being lost on the battlefield. The user must understand the equipment and employ it effectively and within its means.

Section I. MEANS OF COMMUNICATIONS

Each of the several means of communication has its own advantages and disadvantages (Table 11-1, page 11-2).

MESSENGERS

ADVANTAGES

- Messengers are the most secure means of communication.
- Messengers can hand carry large maps with overlays.
- Messengers can deliver supplies along with messages.
- Messengers are flexible (can travel long/short distances by foot or vehicle).

DISADVANTAGES

- Messengers are slow, especially if traveling on foot for a long distance.
- Messengers might be unavailable, depending on manpower requirements (size of element delivering message).
- Messengers can be captured by enemy.

WIRE

ADVANTAGES

- Wire reduces radio net traffic.
- Wire reduces electromagnetic signature.
- Wire is secure and direct.
- Wire can be interfaced with a radio.

DISADVANTAGES

- Wire has to be carried (lots of it).
- Wire must be guarded.
- Wire is time consuming.

Table 11-1. Comparison of communication methods.

Method	Advantages	Disadvantages
Messengers	• Messengers are the most secure means of communication. • Messengers can hand carry large maps with overlays. • Messengers can deliver supplies along with messages. • Messengers are flexible (can travel long/short distances by foot or vehicle).	• Messengers are slow, especially if traveling on foot for a long distance. • Messengers might be unavailable, depending on manpower requirements (size of element delivering message). • Messengers can be captured by enemy.
Wire	• Wire reduces radio net traffic. • Wire reduces electromagnetic signature. • Wire is secure and direct. • Wire can be interfaced with a radio.	• Wire has to be carried (lots of it). • Wire must be guarded. • Wire is time consuming.
Visual Signals	• Visual signals aid in identifying friendly forces. • Visual signals allow transmittal of prearranged messages. • Visual signals are fast. • Visual signals provide immediate feedback.	• Visual signals can be confusing. • Visual signals are visible from far away. • The enemy might see them, too.
Sound	• Sound can be used to attract attention. • Sound can be used to transmit prearranged messages. • Sound can be used to spread alarms. • Everyone can hear it at once. • Sound provides immediate feedback.	• The enemy hears it also. • Sound gives away your position.
Radio	• Radios are the most frequently used means of communication. • Radios are fast. • Radios are light. • Radios can be interfaced with telephone wire.	• Radio is the least secure means of communication. • Radios require batteries. • Radios must be guarded or monitored.

VISUAL SIGNALS

ADVANTAGES

- Visual signals aid in identifying friendly forces.
- Visual signals allow transmittal of prearranged messages.
- Visual signals are fast.
- Visual signals provide immediate feedback.
- Visual signals are visible from far away.

DISADVANTAGES

- Visual signals can be confusing.
- The enemy might see them, too.

SOUND

ADVANTAGES

- Sound can be used to attract attention.
- Sound can be used to transmit prearranged messages.
- Sound can be used to spread alarms.
- Everyone can hear it at once.
- Sound provides immediate feedback.

DISADVANTAGES

- The enemy hears it also.
- Sound gives away your position.

RADIO

ADVANTAGES

- Radios are the most frequently used means of communication.
- Radios are fast.
- Radios are light.
- Radios can be interfaced with telephone wire.

DISADVANTAGES

- Radio is the least secure means of communication.
- Radios require batteries.
- Radios must be guarded or monitored.

Chapter 11

Section II. RADIOTELEPHONE PROCEDURES

Radio, the least secure means of communication, speeds the exchange of messages and helps avoid errors. Proper radio procedures must be used to reduce the enemy's opportunity to hamper radio communications. Each time you talk over a radio, the sound of your voice travels in all directions. The enemy can listen to your radio transmissions while you are communicating with other friendly radio stations. You must always assume that the enemy is listening to get information about you and your unit, or to locate your position to destroy you with artillery fire.

RULES

11-1. Radio procedure rules, listed below, will help you use transmission times efficiently and avoid violations of communications.

- Prior to operation, assure equipment is properly configured. The TM is a good place to begin. Examples of items to check include tuning, power settings, and connections.
- Change frequencies and call signs IAW unit signal operating instructions (SOI).
- Use varied transmission schedules and lengths.
- Use established formats to expedite transmissions such as sending reports.
- Encode messages or use secure voice.
- Clarity of radio communications varies widely, so use the phonetic alphabet and numbers.
- Transmit clear, complete, and concise messages. When possible, write them out beforehand.
- Speak clearly, slowly, and in natural phrases as you enunciate each word. If a receiving operator must write the message, allow time for him to do so.
- Listen before transmitting to avoid interfering with other transmissions.
- Long messages risk becoming garbled and create increased electronic signature. The use of prowords is essential in reducing transmission time and avoiding confusion.
- Minimize transmission time.

TYPES OF NETS

11-2. Stations are grouped into nets according to requirements of the tactical situation. A Net is two or more stations in communications with each other, operating on the same frequency. Nets can be for voice and/or data communications. The types of nets follow:

Command Net (Command and control the unit's maneuver).

Intelligence Net (Communicate enemy information and develop situational awareness).

Operations and Intelligence Net

Administration and Logistics Net (Coordinate sustainment assets).

PRECEDENCE OF REPORTS

Flash (For initial enemy contact reports).

Immediate (Situations which greatly affect the security of national and allied forces).

Priority (Important message over routine traffic).

Routine (All types of messages that are not urgent).

Communications

MESSAGE FORMAT

Heading--A heading consists of the following information:
1. Identity of distant station and self.
2. Transmission instructions (*Relay To, Read Back, Do Not Answer*).
3. Precedence.
4. FROM/TO.

Text--Text is used to--
1. Separate heading from message with *Break*.
2. State reason for message.

Ending--An ending consists of--
Final Instructions (*Correction, I Say Again, More to Follow, Standby, Execute, Wait*).
OVER or OUT (*never use both together*).

COMMON MESSAGES

11-3. Soldiers should know how to prepare and use the *Nine-Line MEDEVAC Request* and the call for fire.

Nine-Line MEDEVAC Request--For a more detailed description (Table 4-3, FM 4-02.2):
Line 1   Location of pickup site.
Line 2   Radio frequency, call sign, and suffix.
Line 3   Number of patients by precedence.
Line 4   Special equipment required.
Line 5   Patient type.
Line 6   Security of pickup sight (wartime).
Line 6   Number and type of wound, injury, or illness (peacetime).
Line 7   Method of marking pickup site.
Line 8   Patient nationality and status.
Line 9   CBRN contamination (wartime).
Line 9   Terrain description (Peacetime).

Call for Fire--The normal call for fire is sent in three parts, each of which has the following six elements. The six elements, detailed in the sequence in which they are transmitted, follow: For a more detailed explanation for calling for fire, see FM 6-30, Chapter 4:
- Observer identification.
- Warning order.
- Target location.
- Target description.
- Method of engagement.
- Method of fire and control.

PROWORDS

11-4. The following paragraphs discuss common, strength, and readability prowords, as well as radio checks:

Common Prowords--Common prowords are those words used on a regular basis while conducting radio operations. They are NOT interchangeable, as the meanings are specific and clear to the receiver. An example is "Say Again" versus "Repeat." "Say Again" means to repeat the last transmission, while "Repeat" refers to fire support, and means to fire the last mission again (Figure 11-1).

Strength and Readability Prowords--Certain strength and readability prowords must be used during radio checks:

Strength Prowords
- Loud.
- Good.
- Weak.
- Very Weak.
- Poor.

Readability Prowords
- Clear.
- Readable.
- Unreadable.
- Distorted.
- With Interference.
- Intermittent.

Radio Checks--Rating signal strength and readability. An example radio check follows:

Radio Check	What is my strength and readability?
Roger	I received your transmission satisfactorily.

PROWORD	MEANING
ALL AFTER	I refer to the entire message that follows...
ALL BEFORE	I refer to the entire message that precedes...
BREAK	I now separate the text from other parts of the message.
CORRECTION	There is an error in this transmission. This will continue with the last word correctly transmitted.
GROUPS	This message contains the number of groups indicated by the numeral following.
I SAY AGAIN	I am repeating transmission or part indicated.
I SPELL	I shall spell the next word phonetically.
MESSAGE	A message that requires recording is about to follow. (Transmitted immediately after the call.) This proword is not used on nets primarily employed for conveying messages. It is intended for use when messages are passed on tactical or reporting net.
MORE TO FOLLOW	Transmitting station has additional traffic for the receiving station.
OUT	This is the end of my transmission to you and no answer is required or expected.
OVER	This is the end of my transmission to you and a response is necessary. Go ahead: transmit.
RADIO CHECK	What is my signal strength and readability, i.e. How do you hear me?
ROGER	I have received your last transmission satisfactorily, radio check is loud and clear.
SAY AGAIN	Repeat all of your last transmission. Followed by identification data means "repeat - (portion indicated)."
THIS IS	This transmission is from the station whose designator immediately follows.
TIME	That which immediately follows is the time or date-time group of the message.
WAIT	I must pause for a few seconds.
WAIT-OUT	I must pause longer than a few seconds.
WILCO	I have received your transmission, understand it, and will comply, to be used only by the addressee. Since the meaning of ROGER is included in that of WILCO, the two prowords are never used together.
WORD AFTER	I refer to the word of the message that follows.
WORD BEFORE	I refer to the word of the message that precedes.

Figure 11-1. Common prowords.

Chapter 11

OPERATION ON A NET

Preliminary Calls
Bulldog 19, this is Bulldog 29. Over.
Bulldog 29, this is Bulldog 19. Over.
Bulldog 19, this is Bulldog 29. Message. Over.
Bulldog 29, this is Bulldog 19. Send your message. Over.

Correction
Bulldog 19, this is Bulldog 29. Convoy Romeo 3, correction: Romeo 4 should arrive 1630z. Over.

Read Back
Bulldog 19, this is Bulldog 29. Read back. Convoy has arrived. Time 1630z. Over.

Say Again
Bulldog 19, this is Bulldog 29. Request a recovery vehicle to grid 329966. Over.
Bulldog 29, this is Bulldog 19. Say again, all before grid. Over.
Bulldog 19, this is Bulldog 29. I say again. Request a recovery vehicle. Over.

Roger versus Wilco
Bulldog 19, this is Bulldog 29. Request a recovery vehicle to grid 329966. Over.
Bulldog 29, this is Bulldog 19. Roger. Over.
Bulldog 19, this is Bulldog 29. MOVE TO GRID 329966. Over.
Bulldog 29, this is Bulldog 19. WILCO. Over.
Bulldog 29, this is Bulldog 19. Roger. Over.

Section III. COMMUNICATIONS SECURITY

Communications security (COMSEC) consists of measures and controls to deny unauthorized persons information from telecommunications and ensure authenticity of such telecommunication. COMSEC material includes--

- Cryptographic security.
- Transmission security.
- Emission security.
- Physical security.

CLASSIFICATIONS

11-5. Classified material, protected against unauthorized access, is information produced and owned by the US Government. Authorized access to (clearance to view) classified material requires a NEED-TO-KNOW designation and the appropriate security clearance. Responsibility to protect security of the material rests with the individual handling it. No person is authorized based solely on their rank, title, or position. They are only authorized by their classification level. The three levels of security classification (clearances) follow:

Top Secret--This classification applies to material that could cause exceptionally grave damage to national security.

Secret--This classification applies to material that could cause serious damage to national security.

Confidential--This classification applies to material that could cause damage to national security.

Note: "For official use only" (FOUO) is a handling instruction, *not* a classification.

Communications

SIGNAL OPERATING INSTRUCTIONS

11-6. The SOI is a COMSEC aid designed to provide transmission security by limiting and impairing enemy intelligence collection efforts. The SOI is a series of orders issued for technical control and coordination of a command or activity. It provides guidance needed to ensure the speed, simplicity, and security of communications.

Types
- Training SOI .... Unclassified or FOUO.
- Operation SOI .... Used only when deployed for mission.
- Exercise SOI .... Used for field training exercises.

Components
- Call signs.
- Frequencies.
- Pyrotechnics.
- Challenge and password.

AUTOMATED NET CONTROL DEVICE

11-7. The ANCD (Figures 11-2 and 11-3) is a handheld device that allows users to store/transmit data via cable and retrieve COMSEC. It is enabled for night viewing. It also features a water resistant case and sufficient backup memory.

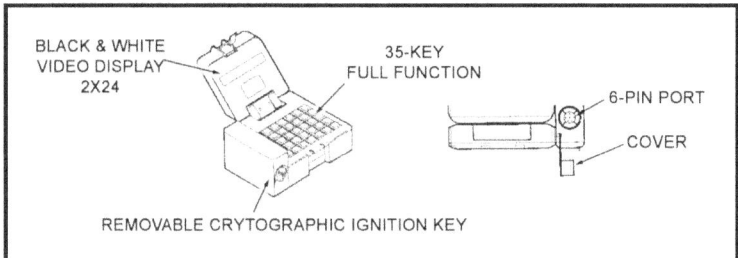

Figure 11-2. Automated net control device.

LAMP	ZERO	MAIN MENU	RECV	SEND	ABORT	ON/OFF
A P UP	B BAT	C CLR	D DELE	E 7	F 8	G 9
H P DN	I ^	J	K	L 4	M 5	N 6
O <	P SPACE	Q >	R	S 1	T 2	U 3
LOCK LTR	V V	W -	X /	Y 0	Z .	ENTER

Figure 11-3. Automated net control device keypad.

Chapter 11

MAIN MENU

11-8. The main menu has three choices:

- SOI pertains to SOI Information.
- Radio pertains to COMSEC keys/FH data to be loaded into the radio.
- Supervisor pertains to areas performed by the supervisor only.

11-9. To select areas of the main menu--

1. Use the ARROW key function by pressing either the left or right arrow keys, and then press the ENTER key.

2. Press the corresponding capital letter on the keyboard to take you directly to a specific topic:
   S Signal operating instructions
   R Radio
   U Supervisor

```
ANCD MAIN MENU
SOI   Radio   sUpervisor
```

CALL SIGNS

11-10. Call signs have two parts (Figure 11-4):

- Designation call sign identifies the major unit (corps, division, brigade, or battalion).
- Suffix and expanders identify individuals by position.

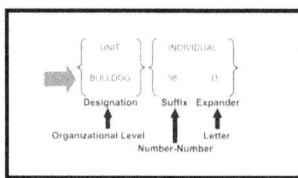

Figure 11-4. Call signs.

TIME PERIODS

11-11. Set times (Figure 11-5).

- Ten time periods in an SOI, which change by calendar day at a designated time (typically 2400Z).
- Each time period has a different call sign and frequency for each unit.
- ANCD breaks them into 2 SETS: TP: 1-5 and TP: 6-10.

Note: Follow unit SOP for preventive and immediate action measures for capture or compromise of SOI or systems.

```
Set:   2BDE      1/4
Edn:   KTV 1234A TP: 1-5

Set:   2BDE      2/4
Edn:   KTV 1234A TP: 6-10
```

Figure 11-5. Time periods.

Section IV. EQUIPMENT

This section discusses radio, wire, and telephone equipment.

RADIOS

11-12. Radios are particularly suited for use when you are on-the-move and need a means of maintaining command and control. Small handheld or backpacked radios that communicate for only short distances are found at squad and platoon level. As the need grows to talk over greater distances and to more units, the size and complexity of radios are increased. The enhancement in modern radio technology is based upon three basic radio systems, each with its own capabilities and characteristics: improved high frequency radio (IHFR), single-channel ground and airborne radio systems (SINCGARS), and single-channel tactical satellite communications (SATCOM). A radio set has a transmitter and receiver. Other items necessary for operation include a battery for a power source, and an antenna for radiation and reception of radio waves. The transmitter contains an oscillator that generates radio frequency (RF) energy in the form of alternating current (AC). A transmission line or cable feeds the RF to the antenna. The antenna converts the AC into electromagnetic energy, which radiates into space. Many radio antennas can be configured or changed to transmit in all directions or in a narrow direction to help minimize the enemy's ability to locate the transmitter. A keying device is used to control the transmission.

AN/PRC-148 MULTIBAND, INTRATEAM RADIO

11-13. See Figure 11-6.

Chapter 11

 Receiver Transmitter Unit (RTU)	
Range	5 kilometers
Antennas	30-90 MHz
	30-512 MHz (reduced gain below 90 MHz)
Batteries	Rechargeable Lithium-Ion (2)
	Nonrechargeable (2) and case
Optimal battery life	10 hrs
Weight	2 pounds
Interoperability	AN/PRC-119 SINCGARS
Transceiver/battery holster	
System carrying bag	

Figure 11-6. AN/PRC-148 multiband intrateam radio (MBITR).

Note: Actual battery life depends upon radio settings, environmental considerations, and battery age.

IC-F43

11-14. The IC-F43 portable UHF transceiver is a two-way, intersquad, land-mobile radio with squad radio voice communications and secure protection (Figure 11-7).

Range	2.5 kilometers (2,500 meters)
Optimal battery life	10 hours
Weight	Less than 1 pound
Interoperability	AN/PRC-119 SINCGARS

Figure 11-7. IC-F43 portable UHF transceiver.

RT 1523A-D (SIP)

11-15. Running the self-test in the system improvement program (SIP) with COMSEC set to PT will produce a FAIL5 message. Change COMSEC to CT to clear the error message (Figure 11-8).

Communications

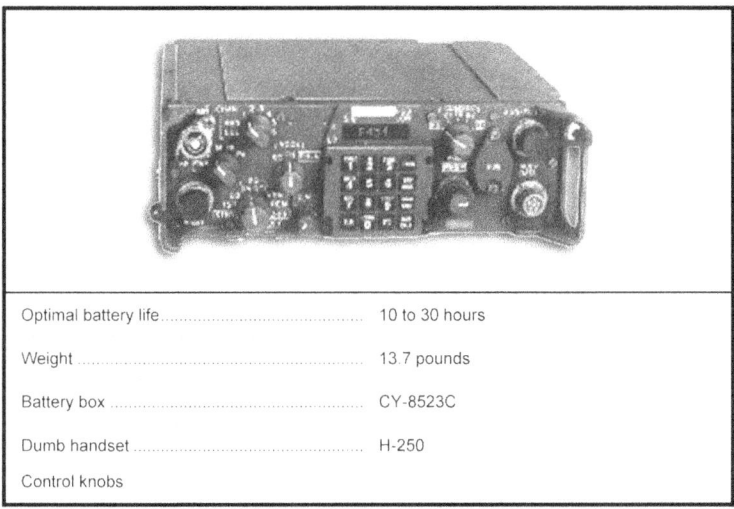

Optimal battery life	10 to 30 hours
Weight	13.7 pounds
Battery box	CY-8523C
Dumb handset	H-250
Control knobs	

Figure 11-8. AN/PRC-119A-D SIP.

RT 1523E

11-16. See Figure 11-9 for more information about the advanced system improvement program (ASIP).

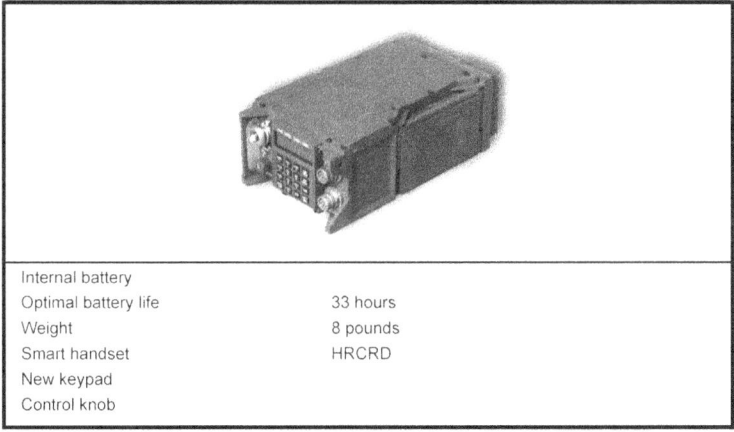

Internal battery	
Optimal battery life	33 hours
Weight	8 pounds
Smart handset	HRCRD
New keypad	
Control knob	

Figure 11-8. AN/PRC-119E advanced system improvement program (ASIP).

WIRE

11-17. The decision to establish wire communications depends on the need; time required and available to install and use; and capability to maintain. The supply of wire on-hand, the expected resupply, and future needs must also be considered. Wire communications can be used in most terrain and tactical situations. When in the defense, units normally communicate by wire and messenger instead of by radio. Your leaders will often have you lay the wire, and install and operate the field phones.

11-18. A *surface line* is field wire laid DR-8 laid on the ground. Lay surface lines loosely with plenty of slack. Slack makes installation and maintenance easier. Surface lines take less time and fewer Soldiers to install. When feasible, dig small trenches for the wire to protect it from shell fragments of artillery or mortar rounds. Conceal wire routes crossing open areas from enemy observation. Tag all wire lines at switchboards, and at road, trail, and rail crossings to identify the lines and make repair easier if a line is cut. An *overhead line* is field wire laid above the ground. Lay overhead lines near command posts, in assembly areas, and along roads where heavy vehicular traffic may drive off the road. Also, lay them at road crossings where trenches cannot be dug, if culverts or bridges are unavailable. Those lines are the least likely to be damaged by vehicles or weather.

TELEPHONE EQUIPMENT

11-19. The telephone set TA-1 is a sound-powered phone with both visual and audible signals. Its range is 4 miles using WD-1 wire. Telephone set TA-312 is a battery-powered phone. Its range is 14 miles using WD-1 wire.

TA-1 TELEPHONE

11-20. To install the TA-1 telephone--

- Strip away half an inch of insulation from each strand of the WD-1 wire line.
- Press the spring-loaded line binding posts and insert one strand of the wire into each post.
- Adjust the signal volume-control knob to LOUD.
- Press the generator lever several times to call the other operator
- Listen for the buzzer sound.
- Turn the buzzer volume-control knob to obtain the desired volume.
- See if the indicator shows four white luminous markings.
- If so, press the push-to-talk switch to reset the visual indicator.

TA-312 TELEPHONE

11-21. To install the TA-312 telephone--

- Strip away 1/2 inch of insulation from each strand of the WD-1 wire line.
- Press the spring-loaded line binding posts and insert one strand of the wire into each post.
- Adjust the buzzer volume-control knob to LOUD.
- Turn the INT-EXT switch to INT.
- Turn the circuit selector switch to LB.
- Insert two BA-30 batteries into the battery compartment (one up and one down).
- Seat the handset firmly in the retaining cradle.
- Turn the hand crank rapidly a few turns.
- Remove the handset from the retaining cradle and wait for the other operator to answer.
- Press the push-to-talk switch to talk.
- Release the push-to-talk switch to listen.

CE-11 REEL

11-22. The CE-11 reel is a lightweight, portable unit used to lay and pick up short wire lines. It includes the RL-39 band cable-reeling machine, axle, crank, carrying handles, straps ST-34 and ST-35, and telephone set TA-1/PT, all of which may be authorized as a unit or listed separately in the TOE. The DR-8 reel cable and the WD-1/TT field wire (400 feet) are always listed separately from the RL-39 and each other.

This page intentionally left blank.

Chapter 12
Survival, Evasion, Resistance, and Escape

Continuous operations and fast-moving battles increase your chances of either becoming temporarily separated from your unit or captured. Whether you are separated from your unit or captured, your top priority should be rejoining your unit or making it to friendly lines. If you do become isolated, every Soldier must continue to fight, evade capture, and regain contact with friendly forces. If captured, detained, or held hostage, individual Soldiers must live, act, and speak in a manner that leaves no doubt they adhere to the traditions and values of the US Army and the Code of Conduct.

SURVIVAL

12-1. The acrostic **SURVIVAL** can help guide your actions in any situation (Figure 12-1 [short list] and Figure 12-2 [explanations]). Learn what each letter represents, and practice applying these guidelines when conducting survival training:

Chapter 12

S	**Size up the Situation (Surroundings, Physical Condition, Equipment).** In combat, conceal yourself from the enemy. Security is key. "Size up" the battlespace (situation, surroundings, physical condition, and equipment). Determine if the enemy is attacking, defending, or withdrawing. Make your survival plan, considering your basic physical needs—water, food, and shelter. *Surroundings*--Figure out what is going on around you and find the rhythm or pattern of your environment. It incl animal and bird noises, and movements and insect sounds. It may also include enemy traffic and civilian movemer *Physical Condition*--The pressure of the previous battle you were in (or the trauma of being in a survival situation) have caused you to overlook wounds you received. Check your wounds and give yourself first aid. Take car prevent further bodily harm. For instance, in any climate, drink plenty of water to prevent dehydration. If you are cold or wet climate, put on additional clothing to prevent hypothermia. *Equipment*--Perhaps in the heat of battle, you lost or damaged some of your equipment. Check to see what equipment have and its condition.
U	**Use All Your Senses: Undue Haste Makes Waste** Evaluate the situation. Note sounds and smells. Note temperature changes. Stay observant and act carefully. An unplan action can result in your capture or death. Avoid moving just to do something. Consider all aspects of your situation befo you do anything. Also, if you act in haste, you might forget or lose some of your equipment. You might also get disoriente and not know which way to go. Plan your moves. Stay ready to move out quickly, but without endangering yourself, if the enemy is near.
R	**Remember Where You Are** Find out who in your group has a map or compass. Find yourself on a map and continually reorient yourself on your loca and destination. Ensure others do the same. Rely on yourself to keep track of your route. This will help you make intellige decisions in a survival or evasion situation. Always try to determine, as a minimum, how your location relates to-- • Enemy units and controlled areas. • Friendly units and controlled areas. • Local water sources (especially important in the desert). • Areas that will provide good cover and concealment.
V	**Vanquish Fear and Panic** Fear and panic are your greatest enemies. Uncontrolled, they destroy the ability to make intelligent decisions, or they car you to react to feelings and imagination rather than the situation. They will drain your energy, and lead to other negative emotions. Control them by remaining self-confident and using what you learned in your survival training.
I	**Improvise** Americans are unused to making do. This can hold you back in a survival situation. Learn to improvise. Take a tool desig for a specific purpose and see how many other uses you can find for it. Learn to use natural objects around you for differ needs, for example, use a rock for a hammer. When your survival kit inevitably wears out, you must use your imaginatior fact, when you can improvise suitable tools, do so, and save your survival kit items for times when you have no such options.
V	**Value Living** When faced with the stresses, inconveniences, and discomforts of a survival situation. Soldiers must maintain a high valu on living. The experience and knowledge you have gained through life and Army training will have a bearing on your will live. Perseverance, a refusal to give in to problems and obstacles that face you, will give you the mental and physical strength to endure.
A	**Act like the Natives** Locals (indigenous people and animals) have already adapted to an environment that is strange to you. • Observe daily routines of local people. Where do they get food and water? When and where do they eat? What time they go to bed and get up? The answers to these questions can help you avoid capture. • Watch animals, who also need food, water, and shelter, to help you find the same. • Remember that animals may react to you, revealing your presence to the enemy. • In friendly areas, gain rapport with locals by showing interest in their customs. Studying them helps you learn to resp them, allows you to make valuable friends, and, most importantly, helps you adapt to their environment. All of these increase your chance of survival.
L	**Live by your Wits, but for Now Learn Basic Skills** Having basic survival and evasion skills will help you live through a combat survival situation. Without these skills, your chance of survival is slight. • Learn these skills now—not en route to, or in, battle. Know the environment you are going into and practice basic ski geared to the environment. Equipping yourself for the environment beforehand will help determine whether you survi For instance, if you are going to a desert, know how to get--and purify--water. • Practice basic survival skills during all training programs and exercises. Survival training reduces fear of the unknowi gives you self-confidence, and teaches you to live by your wits.

Figure 12-1. SURVIVAL.

THREE-PHASE SURVIVAL KIT

12-2. A useful technique for organizing survival is the three-phase individual survival kit. The content of each phase of the kit depends on the environment in the AO and available supplies. An example of the contents of a three-phase survival kit is as follows:

Phase 1 (Extreme)

12-3. Soldier without any equipment (load-bearing equipment or rucksack). Items to be carried (and their suggested uses) include--

- Safety pins in hat (fishing hooks or holding torn clothes).
- Utility knife with magnesium fire starter on 550 cord wrapped around waist (knife, making ropes, and fire starter).
- Wrist compass (navigation).

Phase 2 (Moderate)

12-4. Soldiers with load-bearing equipment. Load-bearing equipment should contain a small survival kit. Kit should be tailored to the AO and should only contain basic health and survival necessities:

- 550 cord, 6 feet (cordage, tie down, fishing line, weapons, and snares).
- Waterproof matches or lighter (fire starter).
- Iodine tablets (water purification, small cuts).
- Fish hooks or lures (fishing).
- Heavy duty knife with sharpener, bayonet type (heavy chopping or cutting).
- Mirror (signaling).
- Tape (utility work).
- Aspirin.
- Clear plastic bag (water purification, solar stills).
- Candles (heat, light).
- Surgical tubing (snares, weapons, drinking tube).
- Tripwire (traps, snares, weapons).
- Dental floss (cordage, fishing line, tie down, traps).
- Upholstery needles (sewing, fish hooks).

Phase 3 (Slight)

12-5. Soldier with load-bearing equipment and rucksack. Rucksack should only contain minimal equipment. The following are some examples:

- Poncho (shelters, gather water such as dew).
- Water purification pump.
- Cordage (550), 20 feet.
- Change of clothes.
- Cold and wet weather jacket and pants.
- Poncho liner or lightweight sleeping bag.

Note: Items chosen for survival kits should have multiple uses. The items in the above list are only suggestions.

Chapter 12

EVASION

12-6. Evasion is the action you take to stay out of the enemy's hands when separated from your unit and in enemy territory. There are several courses of action you can take to avoid capture and rejoin your unit. You may stay in your current position and wait for friendly troops to find you, or you may try to move and find friendly lines. Below are a few guidelines you can follow.

PLANNING

12-7. Planning is essential to achieve successful evasion. Follow these guidelines for successful evasion:

- Keep a positive attitude.
- Use established procedures.
- Follow your EPA (evasion plan of action).
- Be patient.
- Drink water.
- Conserve strength for critical periods.
- Rest and sleep as much as possible.
- Stay out of sight.

ODORS

12-8. Avoid the following odors (they stand out and may give you away):

- Scented soaps and shampoos.
- Shaving cream, after-shave lotion, or other cosmetics.
- Insect repellent (camouflage stick is least scented).
- Gum and candy (smell is strong or sweet).
- Tobacco (odor is unmistakable).
- Mask scent using crushed grasses, berries, dirt, and charcoal.

EVASION PLAN OF ACTION

12-9. Establish--

- Suitable area for recovery.
- Selected area for evasion.
- Neutral or friendly country or area.
- Designated area for recovery.

SHELTERS

12-10. Keep the following guidelines in mind concerning shelters:

- Use camouflage and concealment.
- Locate carefully (BLISS, Figure 12-2).
- Choose an area.
    -- Least likely to be searched (for example drainages, rough terrain) and blends with the environment.
    -- With escape routes (*do not* corner yourself).
    -- With observable approaches.
- Locate entrances and exits in brush and along ridges, ditches, and rocks to keep from forming paths to site.

Survival, Evasion, Resistance, and Escape

- Be wary of flash floods in ravines and canyons.
- Conceal with minimal to no preparation.
- Take the radio direction finding threat into account before transmitting from shelter.
- Ensure overhead concealment.

```
B  Blend
L  Low silhouette
I  Irregular shape
S  Small
S  Secluded location
```

Figure 12-2. Tool for remembering shelter locations.

MOVEMENT

12-11. Remember, a moving object is easy to spot. If travel is necessary--

- Mask with natural cover.
- Stay off ridgelines and use the military crest (2/3 of the way up) of a hill.
- Restrict to periods of low light, bad weather, wind, or reduced enemy activity.
- Avoid silhouetting.
- *Do* the following at irregular intervals:
  -- *Stop* at a point of concealment.
  -- *Look* for signs of human or animal activity (such as smoke, tracks, roads, troops, vehicles, aircraft, wire, and buildings). Watch for trip wires or booby traps, and avoid leaving evidence of travel. Peripheral vision is more effective for recognizing movement at night and twilight.
  -- *Listen* for vehicles, troops, aircraft, weapons, animals, and so forth.
  -- *Smell* for vehicles, troops, animals, fires, and so forth.
- Use noise discipline; check clothing and equipment for items that could make noise during movement and secure them.
- Break up the human shape or recognizable lines.
- Camouflage evidence of travel. Route selection requires detailed planning and special techniques (irregular route/zigzag).
- Concealing evidence of travel. Using techniques such as:
- Avoid disturbing vegetation.
- *Do not* break branches, leaves, or grass. Use a walking stick to part vegetation and push it back to its original position.
- *Do not* grab small trees or brush. (This may scuff the bark or create movement that is easily spotted. In snow country, this creates a path of snow-less vegetation revealing your route.)
- Pick firm footing (carefully place the foot lightly but squarely on the surface to avoid slipping).
- Try not to--
  -- Overturn ground cover, rocks, and sticks.
  -- Scuff bark on logs and sticks.
  -- Make noise by breaking sticks. (Cloth wrapped around feet helps muffle noise.)
  -- Mangle grass and bushes that normally spring back.
- Mask unavoidable tracks in soft footing.
  -- Place tracks in the shadows of vegetation, downed logs, and snowdrifts.

- -- Move before and during precipitation, allows tracks to fill in.
- -- Travel during windy periods.
- -- Take advantage of solid surfaces (such as logs and rocks) leaving less evidence of travel.
- -- Tie cloth or vegetation to feet, or pat out tracks lightly to speed their breakdown or make them look old.
- Secure trash or loose equipment and hide or bury discarded items. (Trash or lost equipment identifies who lost it.)
- If pursued by dogs, concentrate on defeating the handler.
  - -- Travel downwind of dog/handler, if possible.
  - -- Travel over rough terrain and/or through dense vegetation to slow the handler.
  - -- Travel downstream through fast moving water.
  - -- Zigzag route if possible, consider loop-backs and "J" hooks.
- Penetrate obstacles as follows:
  - -- Enter deep ditches feet first to avoid injury.
  - -- Go around chain-link and wire fences. Go under fence if unavoidable, crossing at damaged areas. *Do not* touch fence; look for electrical insulators or security devices.
  - -- Penetrate rail fences, passing under or between lower rails. If this is impractical, go over the top, presenting as low a silhouette as possible.
  - -- Cross roads after observation from concealment to determine enemy activity. Cross at points offering concealment such as bushes, shadows, or bends in the road. Cross in a manner leaving footprints parallel (cross step sideways) to the road.

RESISTANCE

12-12. Figure 12-3 shows the Code of Conduct, which prescribes how every Soldier of the US armed forces must conduct himself when captured (or faced with the possibility of capture).

Survival, Evasion, Resistance, and Escape

> I. I am an American, fighting in the forces which guard my country and our way of life. I am prepared to give my life in their defense.
>
> II. I will never surrender of my own free will. If in command, I will never surrender the members of my command while they still have the means to resist.
>
> III. If I am captured, I will continue to resist by all means available. I will make every effort to escape and aid others to escape. I will accept neither parole nor special favors from the enemy.
>
> IV. If I become a prisoner of war, I will keep faith with my fellow prisoners. I will give no information or take part in any action which might be harmful to my comrades. If I am senior, I will take command. If not, I will obey the lawful orders of those appointed over me, and will back them up in every way.
>
> V. When questioned, should I become a prisoner of war, I am required to give only name, rank, service number, and date of birth. I will evade answering further questions to the utmost of my ability. I will make no oral or written statements disloyal to my country (and its allies) or harmful to their cause.
>
> VI. I will never forget that I am an American, fighting for freedom, responsible for my actions, and dedicated to the principles that made my country free. I will trust in my God and in the United States of America.

Figure 12-3. Code of Conduct.

Article I--Soldiers have a duty to support US interests and oppose US enemies regardless of the circumstances, whether located in a combat environment or OOTW (operations other than war) resulting in captivity, detention, or a hostage situation. Past experience of captured Americans reveals that honorable survival in captivity requires that the Soldier possess a high degree of dedication and motivation. Maintaining these qualities requires knowledge of, and a strong belief in, the following:

- The advantages of American democratic institutions and concepts.
- Love and faith in the US and a conviction that the US cause is just.
- Faith and loyalty to fellow POWs.

Article II--Members of the Armed Forces may never surrender voluntarily. Even when isolated and no longer able to inflict casualties on the enemy or otherwise defend themselves, Soldiers must try to evade capture and rejoin the nearest friendly force. Surrender is the willful act of members of the Armed Forces turning themselves over to enemy forces when not required by utmost necessity or extremity. Surrender is always dishonorable and never allowed. When there is no chance for meaningful resistance, evasion is impossible, and further fighting would lead to death with no significant loss to the enemy. Members of Armed Forces should view themselves as "captured" against their will, versus a circumstance that is seen as voluntarily "surrendering." Soldiers must remember the capture was dictated by the futility of the situation and overwhelming enemy strengths.

Article III--The misfortune of capture does not lessen the duty of a member of the Armed Forces to continue resisting enemy exploitation by all means available. Contrary to the Geneva Conventions, enemies whom US forces have engaged since 1949 have regarded the POW compound as an extension of the battlefield. The POW must be prepared for this fact. In the past, enemies of the US have used physical and mental harassment, general mistreatment, torture, medical neglect, and political indoctrination against POWs. POWs must not seek special privileges or accept special favors at the expense of fellow POWs. POWs must be prepared to take advantage of escape

Chapter 12

opportunities whenever they arise. The US does not authorize any Military Service member to sign or enter into any such parole agreement.

Article IV--Officers and NCOs shall continue to carry out their responsibilities and exercise their authority in captivity. Informing on, or any other action detrimental to a fellow POW, is despicable and expressly forbidden. POWs especially must avoid helping the enemy to identify fellow POWs who may have knowledge of value to the enemy and who may be made to suffer coercive interrogation. Strong leadership is essential to discipline. Without discipline, camp organization, resistance, and even survival may be impossible. Personal hygiene, camp sanitation, and care of the sick and wounded are imperative. Wherever located, POWs should organize in a military manner under the senior military POW eligible for command. The senior POW (whether officer or enlisted) in the POW camp or among a group of POWs shall assume command according to rank without regard to Military Service.

Article V--When questioned, a POW is required by the Geneva Conventions and the Code of Conduct, and is permitted by the Uniform Code of Military Justice (UCMJ), to give name, rank, service number, and date of birth. The enemy has no right to try to force a POW to provide any additional information. However, it is unrealistic to expect a POW to remain confined for years reciting only name, rank, service number, and date of birth. If a POW finds that, under intense coercion, he unwillingly or accidentally discloses unauthorized information, the Service member should attempt to recover and resist with a fresh line of mental defense. The best way for a POW to resist is to keep faith with the US, fellow POWs, and oneself to provide the enemy with as little information as possible.

Article VI--A member of the Armed Forces remains responsible for personal actions at all times. Article VI is designed to assist members of the Armed Forces to fulfill their responsibilities and survive captivity with honor. The Code of Conduct does not conflict with the UCMJ, which continues to apply to each military member during captivity or other hostile detention. Failure to adhere to the Code of Conduct may subject Service members to applicable disposition under the UCMJ. A member of the Armed Forces who is captured has a continuing obligation to resist all attempts at indoctrination and must remain loyal to the US.

ESCAPE

12-13. Escape is the action you take to get away from the enemy if you are captured. The best time for escape is right after capture as you will be in a better physical and mental condition. Bad food and bad treatment during capture add to the already stressful fact of captivity. When detained, you will be given minimal rations that are barely enough to sustain life and certainly not enough to build up a reserve of energy. The physical treatment, medical care, and rations of prison life quickly cause physical weakness, night blindness, and loss of coordination and reasoning power. Once you have escaped, it may not be easy to contact friendly troops or get back to their lines, even when you know where they are. Learn and use the information in this chapter to increase your chance of survival on today's battlefield. For more information, see FM 3-05.70 and ATP 3-50.3. Other reasons for escaping early include--

- Friendly fire or air strikes may cause enough confusion and disorder to provide a chance to escape.
- The first guards usually have less training in handling prisoners than the next set. You have a better chance of getting away from the first ones.
- You might know something about the area where you were first captured. You might even know the locations of nearby friendly units.
- The way you escape depends on what you can think of to fit the situation.
- The only general rules are to escape early and when the enemy is distracted.

Chapter 13

Chemical, Biological, Radiological, or Nuclear Weapons

The threat or use of chemical, biological, radiological, and nuclear (CBRN) is a possible condition of future warfare. You could encounter chemical and biological (CB) weapons in the early stages of war to disrupt US operations and logistics. In many regions where the US is likely to deploy forces, potential adversaries may use CB weapons. To meet this challenge, you must be properly trained and equipped to operate effectively and decisively in the face of any CBRN attacks.

Section I. CHEMICAL WEAPONS

A toxic chemical agent is any toxic chemical that, through its chemical action on life processes, can cause death, temporary incapacitation, or permanent harm to humans. Additionally, you can experience significant physiological effects. Chemical agents, further divided into chemical warfare (CW) agents, are classified according to their physical states, physiological actions, and uses. Chemical agents may appear in the field in the forms of vapors, aerosols, or liquids. The terms persistent and nonpersistent describe the time chemical agents remain in an area and do not classify the agents technically. The persistent chemical agents may last anywhere from hours up to days and will necessitate future decontamination and the wearing of protective equipment in that area; whereas nonpersistent chemical agents will last for only a matter of minutes to hours, but are usually more lethal. Chemical agents having military significance are categorized as nerve, blister, blood, incapacitating, or choking agents.

TYPES

13-1. These chemical agents kill, seriously injure, or incapacitate unprotected personnel when employed as discussed in the following paragraphs.

NERVE AGENTS

13-2. If you are in a contaminated area, protective clothing and a mask are the only sufficient protection against nerve agents. Nerve agents act quickly; effects can occur seconds, minutes, or hours after exposure.

Symptoms

- Difficulty breathing.
- Drooling.
- Nausea.
- Vomiting.
- Convulsions.
- Dim vision (sometimes).

Convulsant Antidote for Nerve Agents

13-3. The CANA is an autoinjector containing 10 milligrams of diazepam for intramuscular administration to control nerve agent induced seizures (Figure 13-1). Administration of atropine and 2

Chapter 13

PAM alone often does not prevent the occurrence of severe and long lasting convulsions after nerve agent exposure. The CANA, designed for buddy aid administration and not self administration, is intended to terminate the convulsions.

Figure 13-1. CANA.

CAUTION

Use the casualty's CANA when providing aid. Do not use your own. If you do, you might not have any antidote available when *you* need it.

Antidote Treatment, Nerve Agent, Autoinjector System--When Mark I supplies are exhausted, use the antidote treatment, nerve agent, autoinjector (ATNAA) system. A single ATNAA delivers both atropine and two Pam Cl. See FM 4-25.11 and FM 4-02.285 for more information.

Self-Aid--If you have symptoms of nerve agent poisoning, inject one NAAK in your thigh. If symptoms persist, inject a second one. Allow at least 15 minutes between injections, and do not exceed three NAAK injections. First aid measures include (Figure 13-2)--

-- Atropine.
-- Two PAM nerve agent antidote kits (NAAKs).
-- Mark I.
-- Artificial respiration.
-- Protective clothing.
-- Mask.

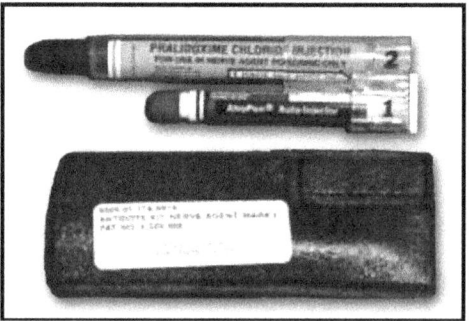

Figure 13-2. NAAK, Mark I.

Buddy-Aid--If a Soldier is so incapacitated that he cannot treat himself, then a buddy must inject three NAAKs at once, IAW the instructions in the kit, and without waiting 15 minutes between injections. Then, he immediately gives artificial respiration. (The maximum number of NAAK injections you may receive is three.) If that does not help, he must administer Mark I. Then, he masks the casualty and quickly seeks medical aid for him. Administer the CANA with the third Mark I to prevent convulsions. Do not administer more than one CANA.

> **WARNING**
>
> The maximum number of NAAK injections is three. Do not exceed this amount. Giving yourself a second set of injections may create a nerve agent antidote overdose, which could cause incapacitation.

> **CAUTION**
>
> Use the casualty's own antidote autoinjectors when providing aid. Do not use your injectors on a casualty. If you do, you might not have any antidote available when needed for self aid.

BLISTER AGENTS

13-4. The symptoms of blister agent poisoning are burning sensations in the skin, eyes, and nose. The symptoms may be immediate or delayed for several hours or days, depending on the type of agent used. If blister agents come in contact with the eyes or skin, decontaminate the areas at once. Decontaminate the eyes by flushing them repeatedly with plain water. If burns or blisters develop on the skin, cover them with sterile gauze or a clean cloth to prevent infection. If you are in a contaminated area, your protective clothing and mask are the only sufficient protection. Seek medical aid quickly.

BLOOD AGENTS

13-5. The symptoms of blood agent poisoning are nausea, dizziness, throbbing headache, red or pink skin/lips, convulsions, and coma. Blood agents cause immediate casualties when absorbed by breathing and inhibits the red blood cells ability to deliver oxygen to the body's organs and tissue. Breathing may become difficult or stop, and death will usually occur within 15 minutes. If you are in a contaminated area, your protective mask is the only sufficient protection. Seek medical aid quickly.

CHOKING AGENTS

13-6. The symptoms of choking agent poisoning are coughing, choking, tightness of the chest, nausea, headache, dry throat, and watering of the eyes. Lungs fill with liquid known as dry land drowning and death results from the lack of oxygen. If you are in a contaminated area, your protective mask is the only sufficient protection. If you have these symptoms immediately, seek medical aid.

INCAPACITATING AGENTS

13-7. Incapacitating agents differ from other chemical agents in that the lethal dose is theoretically many times greater than other agents are. That is, it takes much more of an incapacitating agent to kill someone (FM 3-11.4).

Chapter 13

TOXIC INDUSTRIAL MATERIALS

13-8. See FM 4-02.7, FM 3-11.4, ATP 3-11.37, and TRADOC G2 Handbook No. 1.04 for more information.

Characteristics

- TIMs are produced to prescribed toxicity levels.
- TIMs are administered through inhalation (mostly), ingestion, or absorption.
- MOPP gear may or may not protect against TIMs and other vapor or contact hazards.
- TIMs may be stored or used in any environment and for any tactical purpose, including industrial, commercial, medical, military, or domestic purposes.

Forms

13-9. TIMs may take any of these three forms:

- Chemical (toxic industrial chemicals, or TIC).
- Biological (toxic industrial biological, or TIB).
- Radioactive (toxic industrial radiological, or TIR).

Examples

13-10. Some examples of TIM include--

- Fuels.
- Oils.
- Pesticides.
- Radiation sources.
- Arsenic.
- Cyanide.
- Metals such as mercury and thallium.
- Phosgene.

DETECTION

13-11. Your senses may be unable to detect chemicals. Most agents are odorless, colorless, tasteless, and invisible in battlefield concentrations. However, by using unit-level chemical agent alarms and detection kits, you can detect chemical agents yourself.

M22 AUTOMATIC CHEMICAL AGENT DETECTION ALARM

13-12. The M22 ACADA detects and warns of the presence of standard blister and nerve agents (Figure 13-3). The M22 ACADA system is man-portable, operates independently after system startup, and produces an audible and visual alarm. The M22 system also has a communications interface that automatically provides battlefield warning and reporting. The system monitors the air in all environmental conditions, within standard wheeled and tracked vehicles, and within collective protection shelters.

Chemical, Biological, Radiological, or Nuclear Weapons

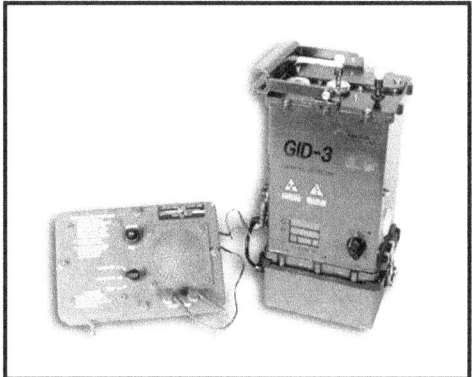

Figure 13-3. M22 ACADA.

IMPROVED CHEMICAL AGENT MONITOR

13-13. The ICAM identifies nerve and blister agent contamination on personnel and equipment (Figure 13-4). The ICAM is a handheld, individually operated, post-attack tool for monitoring chemical agent contamination on personnel and equipment. It detects and discriminates between vapors of nerve and mustard agents. The ICAM gives you instant feedback of chemical hazard levels, and quickly identifies the presence of contamination.

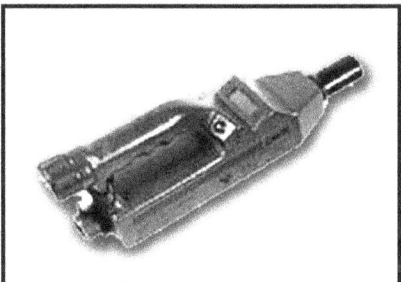

Figure 13-4. M22 ICAM.

M8 CHEMICAL AGENT DETECTOR PAPER

13-14. The M8 detector paper is the only means of identifying the type of chemical agent present in liquid form on the battlefield (Figure 13-5). You will carry one booklet of 25 sheets of M8 paper in the interior pocket of the protective mask carrier. Once you encounter an unknown liquid suspected of being a chemical agent, you must don and check your mask within 9 seconds, and then quickly don the attached hood. Next, alert others who are nearby and don all the rest of your chemical protective clothing. Remove the booklet of M8 paper from your mask carrier, tear a half sheet from the booklet, and, if you can, affix the sheet to a stick. Using the stick as a handle, blot the paper onto the unknown liquid and wait 30 seconds for a color change. To identify the type of agent, compare the resulting color to those on the inside front cover of the booklet. The paper sheets turn dark green, yellow, or red on contact with liquid V-type nerve agents, G-type nerve agents, or blister (mustard) agents. Unfortunately, they cannot detect vapors.

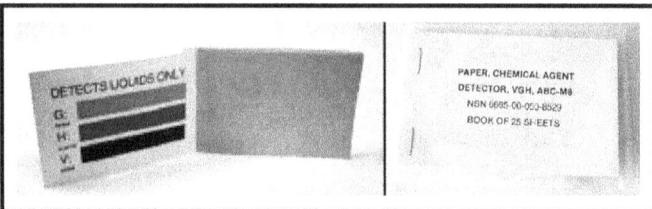

Figure 13-5. M8 chemical agent detector paper.

M9 CHEMICAL AGENT DETECTOR PAPER

13-15. M9 chemical agent detector paper is the most widely used tool used to detect liquid chemical agents (Figure 13-6). M9 paper contains a suspension of an agent-sensitive, red indicator dye in a paper base. It detects and turns pink, red, reddish brown, or red purple when exposed to liquid nerve and blister agents, but it cannot identify the specific agent. Confirm the results of the M9 paper by using the M256 kit. Carry one 30-feet long by 2-inch wide roll of M9 paper with adhesive backing. This will make it easier to wrap a strip of the paper around a sleeve and trouser leg of your protective overgarment. Place the M9 detector paper on opposite sides of your body. If you are right handed, place a strip of M9 paper around your right upper arm, left wrist, and right ankle, and vice versa if you are left handed. You should also attach M9 paper to large pieces of equipment such as shelters or vehicles.

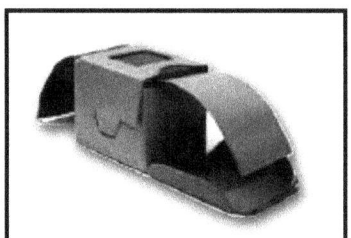

Figure 13-6. M9 chemical agent detector paper.

CAUTION

When attaching M9 chemical paper to equipment, first place the equipment in an area free from dirt, grease, and oil. This is especially important since petroleum products will discolor the paper.

M256 CHEMICAL AGENT DETECTOR KIT

13-16. The M256 has a carrying case, a booklet of M8 paper, 12 disposable sampler-detectors individually sealed in a plastic laminated foil envelope, and a set of instruction cards attached by lanyard to the plastic carrying case (Figure 13-7). The case has a nylon carrying strap and belt attachment. Use this kit to detect and identify blood, blister, and nerve agents, in liquid and vapor forms. You can use it to determine when you can safely unmask, to locate and identify chemical hazards during reconnaissance, and to monitor decontamination effectiveness. Each sampler-detector has a square, impregnated spot for blister agents, a round test spot for blood agents, a star test spot for nerve agents, a lewisite-detecting tablet, and a rubbing tab.

13-17. Of the eight glass ampoules, six contain reagents for testing, and an attached chemical heater contains the other two. When you crush the ampoules between your fingers, formed channels in the plastic sheets direct the flow of liquid reagent, wetting the test spots. Each test spot or detecting tablet develops a distinctive color to show whether a chemical agent is present in the air.

13-18. Follow the directions on the foil packets or in the instruction booklet, and in about 20 minutes, you can conduct a complete test using the liquid-sensitive M8 paper and the vapor-sensitive sampler-detector.

Note: The M256 is not an alarm. It is just a tool for Soldiers to use after they receive *other* warnings about the possible presence of chemical agents.

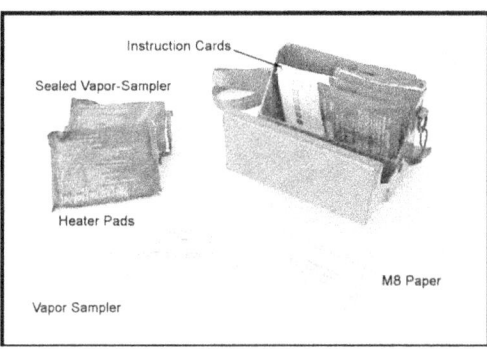

Figure 13-7. M256 chemical agent detector kit.

Chapter 13

PROTECTIVE ACTIONS

13-19. Take these steps to protect against a chemical attack:

- Identify automatic masking criteria.
- Don your protective mask when there is a high probability of a chemical attack, such as when--
    -- A chemical alarm sounds.
    -- A positive reading is obtained on detector paper.
    -- Individuals exhibit symptoms of CB agent poisoning.
    -- You observe a contamination marker.
    -- Your leader tells you to mask.
    -- You see personnel wearing protective masks.
- Respond to the commander's policy of automatic masking.

Note: When chemical weapons have been employed, commanders may modify policy by designating additional events as automatic masking criteria.

- Don, clear, and check your assigned protective mask to protect yourself from CB contamination.
- Give the alarm by yelling "*Gas*" and giving the appropriate hand and arm signal.
- Take cover to reduce exposure, using whatever means are readily available.
- Decontaminate exposed skin using the individual decontaminating kit, as necessary.
- Assume MOPP4.
    -- Cover all skin (head and shoulders already protected by mask and overgarment).
    -- Put on the gloves with liners.
    -- Zip and fasten the overgarment jacket.
    -- Secure the hood, and then secure the overgarment to increase protection.
    -- Put on the overboots.

PROTECTIVE EQUIPMENT

MASK

13-20. Your main protection against a CB attack is your protective mask. The M40A1/A2 series mask provides respiratory, eye, and face protection against CB agents, radioactive fallout particles, and battlefield contaminants. The M42A2 combat vehicle crew (CVC) CB mask has the same components as the M40 A1/A2 (Figure 13-8). In addition, the M42A2 CVC mask has a detachable microphone for wire communications. The canister on the M42A2 mask is attached to the end of a hose and has an adapter for connection to a gas particulate filter unit (GPFU).

Chemical, Biological, Radiological, or Nuclear Weapons

Figure 13-8. Protective mask M40A1/A2 and M42A2 CVC.

PROTECTIVE CLOTHING

13-21. Additionally, protective clothing will provide protection from liquid agents. Protective clothing includes the chemical protective suit, boots, gloves, and helmet cover. The protective clothing is referred to as the joint service lightweight integrated suit technology (JSLIST), which has replaced the battle dress overgarment (BDO). The JSLIST has a service life of 120 days of which 45 days is the maximum wear time. The JSLIST service life begins when the garment is removed from the factory vacuum sealed bag. It can be laundered up to six times for personal hygiene purposes and provides 24 hours of protection against liquid, solid, and/or vapor CB attacks. It also provides protection against radioactive alpha and beta particles.

CHEMICAL PROTECTIVE GLOVES

13-22. The chemical protective gloves protect you against CB agents and alpha and beta radioactive particles as long as they remain serviceable. The glove sets come in three levels of thicknesses (7, 14, and 25 mil). The 7 mil glove set is used for tasks that require extreme sensitivity and will not expose the gloves to harsh treatment. The 14 mil glove set is used by personnel such as vehicle mechanics and weapon crews whose tasks require tactility and will not expose the gloves to harsh treatment. The more durable, 25 mil glove set is used by personnel who perform close combat tasks and other types of heavy labor. If the 14 and 25 mil glove sets become contaminated with liquid chemical agents, decontaminate or replace them within 24 hours after exposure. If the 7 mil glove set becomes contaminated, replace or decontaminate them within 6 hours after exposure.

CHEMICAL PROTECTIVE FOOTWEAR

13-23. Chemical protective footwear includes the green vinyl overshoe (GVO), black vinyl overshoe (BVO), and multipurpose lightweight overboot (MULO). The GVO is a plain, olive drab (OD) green, vinyl overshoe with elastic fasteners. The BVO is very similar to the GVO, except for the color and enlarged tabs on each elastic fastener. You can wear the GVO, BVO, or MULO over your combat boots to protect your feet from contamination by all known agents, vectors, and radioactive alpha and beta particles for a maximum of 60 days of durability and 24 hours of protection against CB agents.

Chemical Protective Footwear Cover

13-24. The chemical protective footwear cover (CPFC) is impermeable and protects your feet from CB agents, vectors, and radioactive alpha and beta particles for a minimum of 24 hours, as long as it remains serviceable.

Chemical Protective Helmet Cover

13-25. The chemical protective helmet cover (CPHC) is a one piece configuration made of butyl coated nylon cloth and gathered at the opening by elastic webbing enclosed in the hem. The cover comes in one size and is OD green color. The helmet cover protects your helmet from CB contamination and radioactive alpha and beta particles.

MISSION-ORIENTED PROTECTIVE POSTURE

13-26. Mission oriented protective posture (MOPP) is a flexible system of protection against chemical agents. Your leader will specify the level of MOPP based on the chemical threat, work rate, and temperature prior to performing a mission. Later, he may direct a change in MOPP according to the changing situation. The MOPP level determines what equipment you must wear and carry. Your unit may increase this level as necessary, but they may not decrease it. The standard MOPP levels are shown in the following chart.

Table 13-1. MOPP levels.

Equipment	MOPP Ready	MOPP0	MOPP1	MOPP2	MOPP3	MOPP4	Mask Only
Mask	Carried	Carried	Carried	Carried	Worn	Worn	Worn***
JSLIST	Ready*	Avail**	Worn	Worn	Worn	Worn	
Overboots	Ready*	Avail**	Avail**	Worn	Worn	Worn	
Gloves	Ready*	Avail**	Avail**	Avail**	Avail**	Worn	
Helmet Cover	Ready*	Avail**	Avail**	Worn	Worn	Worn	

* Item must be available to Soldier within two hours, with replacement available within six hours.
** Item must be positioned within arm's reach of the Soldier.
*** Soldier Never "mask only" if a nerve or blister agent has been used in the AO.

DECONTAMINATION

13-27. Contamination forces your unit into protective equipment that degrades performance of individual and collective tasks. Decontamination restores combat power and reduces casualties that may result from exposure, allowing your unit to sustain combat operations.

Principles--The four principles of decontamination operations are--

1. Decontaminate as soon as possible.
2. Decontaminate only what is necessary.

Chemical, Biological, Radiological, or Nuclear Weapons

3. Decontaminate as far forward as possible.
4. Decontaminate by priority.

LEVELS AND TECHNIQUES--

13-28. The following decontamination levels and techniques are shown in tabular format in Table 13-2 (FM 3-11.4):

1. Immediate decontamination is a basic Soldier survival skill and is performed IAW STP 21-1-SMCT. Personal wipedown removes contamination from exposed skin and individual equipment.
2. Operational decontamination involves MOPP gear exchange and vehicle spraydown. When a thorough decontamination cannot be performed, MOPP gear exchange should be performed within six hours of contamination.
3. Thorough decontamination involves detailed troop decontamination (DTD) and detailed equipment decontamination (DED). Thorough decontamination is normally conducted by company size elements as part of restoration or during breaks in combat operations. These operations require support from a chemical decontamination platoon and a water source or supply.
4. Clearance decontamination provides decontamination to a level that allows unrestricted transport, maintenance, employment, and disposal.

Table 13-2. Decontamination levels and techniques.

Levels	Techniques[1]	Purpose	Best Start Time	Performed By
Immediate	Skin decontamination Personal wipe down Operator wipe down Spot decontamination	Saves lives Stops agent from penetrating Limits agent spread Limits agent spread	Before 1 minute Within 15 minutes Within 15 minutes Within 15 minutes	Individual Individual or buddy Individual or crew Individual or crew
Operational	MOPP gear exchange 2 Vehicle wash down	Provides temporary relief from MOPP4 Limits agent spread	Within 6 hours Within 1 hour (CARC) or within 6 hours (non-CARC)	Unit Battalion crew or decontamination platoon
Thorough	DED and DAD DTD	Provides probability of long-term MOPP reduction	When mission allows reconstitution	Decontamination platoon contaminated unit
Clearance	Unrestricted use of resources	METT-TC depending on the type of equipment contaminated	When mission permits	Supporting strategic resources
The techniques become less effective the longer they are delayed. Performance degradation and risk assessment must be considered when exceeding 6 hours.				

DECONTAMINATING KITS

M291 Skin Decontaminating Kit--The M291 Skin Decontamination Kit has a wallet like carrying pouch containing six individual decontamination packets, enough to perform three complete skin decontaminations (Figure 13-9). Instructions for use are marked on the case and packets. Each packet contains an applicator pad filled with decontamination powder. Each pad provides you with a single step, nontoxic/nonirritating decontamination application. Decontamination is accomplished by application of a black decontamination powder contained in the applicator pad. The M291 can be used on the skin, including the face and around wounds as well as some personal equipment such as your rifle, mask, and gloves. The M291 allows you to completely decontaminate yourself and equipment skin through physical removal, absorption, and neutralization of toxic agent with no long term harmful effects.

Chemical, Biological, Radiological, or Nuclear Weapons

> **WARNING**
>
> This kit is for external use only. It may be slightly irritating to eyes or skin. Be sure to keep the decontamination powder out of eyes, cuts, or wounds, and avoid inhalation of the powder.

M295 Equipment Decontamination Kit--The M295 is issued in boxes of 20 kits (Figure 13-10). The M295 kit has a pouch containing four individual wipe down mitts; each enclosed in a soft, protective packet. The packet is designed to fit comfortably within a pocket of the JSLIST. The M295 allows you to decontaminate through the physical removal and absorption of chemical agents. Each individual wipe down mitt in the kit is comprised of a decontaminating powder contained within a pad material and a polyethylene film backing. In use, powder from the mitt is allowed to flow freely through the pad material. Decontamination is accomplished by adsorption of the liquid agent by the resin and the pad.

M100 Sorbent Decontamination System--The M100 Sorbent Decontamination System (SDS) replaces the M11 and M13 decontamination apparatuses portable (DAPs) and the DS2 used in operator spraydown (immediate decontamination) with a sorbent powder (Figure 13-11). The M100 SDS uses a reactive sorbent powder to remove and neutralize chemical agent from surfaces. Use of the M100 SDS decreases decontamination time and eliminates the need for water. Decontaminate key weapons with M100 SDS or slurry. After decontamination, disassemble weapons and wash, rinse, and oil them to prevent corrosion. Decontaminate ammunition with M100 SDS, wipe with gasoline soaked rags, and then dry it. If M100 SDS is unavailable, wash ammunition in cool, soapy water, then dry it thoroughly. Decontaminate optical instruments by blotting them with rags, wiping with lens cleaning solvent, and then letting them dry. Decontaminate communication equipment by airing, weathering, or hot air (if available). The M100 SDS includes--

- Two 0.7 lb packs of reactive sorbent powder
- Two wash mitt-type sorbent applicators
- Case
- Straps
- Detailed instructions.
- Chemical agent-resistant mounting bracket (optional).

Figure 13-9.
M291 skin
decontaminating kit.

Figure 13-10.
M295 equipment
decontamination kit.

Figure 13-11.
M100 SORBENT
Decontamination System.

Chapter 13

Section II. BIOLOGICAL WEAPONS

Biological agents are microorganisms that cause disease in personnel, plants, or animals or cause the deterioration of material. They create a disease hazard where none exists naturally. They may be dispersed as sprays by generators, or delivered by explosives, bomblets, missiles, and aircraft. They may also be spread by the release of germ carrying flies, mosquitoes, fleas, and ticks. Biological agents can be classified as toxins or pathogens.

TYPES

13-29. Biological agents are classified as toxins or pathogens.

Toxins--Toxins are biologically derived poisonous substances produced as by products of microorganisms, fungi, plants, or animals. They can be naturally or synthetically produced.

Pathogens--Pathogens are infectious agents that cause disease in man, animals, or plants and include bacteria, viruses, rickettsias, protests, fungi, prions or other biologically derived products.

DETECTION

13-30. Biological agents may be disseminated as aerosols, liquid droplets, or dry powder. Attacks with biological agents can be very subtle or direct, if favorable weather conditions prevail. In nearly all circumstances, you will not know a biological attack has occurred. Symptoms can appear from minutes to days after an attack has occurred. Indicators may include -

- Mysterious illness (many individuals sick for unknown reasons).
- Large numbers of vectors, such as insects or unusual insects.
- Large numbers of dead or strange acting (wild and domestic) animals.
- Mass casualties with flu like symptoms—fever, sore throat, skin rash, mental abnormalities, pneumonia, diarrhea, dysentery, hemorrhaging, or jaundice.
- Artillery shells with less powerful explosions than HE rounds.
- Aerial bombs that pop rather than explode.
- Mist or fog sprayed by aircraft or aerosol generators.
- Unexploded bomblets found in the area.

DECONTAMINATION

13-31. Soldiers can use household bleach for biological decontamination.

- Dilute for use on equipment.
- Apply undiluted in the general area of contamination.

PROTECTION

13-32. If threat forces attack with biological agents, you may have little--if any--warning. When a high probability of an attack exists, your unit might assume MOPP4 to protect against contamination. MOPP gear generally protects you against biological agents, but an agent can gain entry through openings such as buttonholes; zippers; stitches; poor seal at ankles, wrists, and neck; or through minute pores in the clothing fabric. Some toxins, however, require the same amount of protection as chemical agents. Consider any unknown agent cloud as a sign of a biological attack and take the same actions prescribed for a chemical attack. Protective measures can be accomplished long before a biological attack happens as follows:

Up-to-Date Immunizations

13-33. Immunizations reduce the chances of becoming biological casualties. Proper immunizations protect against many known disease producing biological agents. You should receive basic immunizations. Medical personnel will periodically screen your records and keep them up to date. When your unit deploys to areas where specific diseases are prevalent, readiness preparation may include providing additional immunizations for needed protection.

Good Hygiene

13-34. The best defense against biological agents is good personal hygiene, which means keeping your body as clean as possible. This not only means washing your face and hands, but all parts of your body—particularly your feet and exposed skin. Shaving may seem unimportant in the field, but it is necessary to achieve a proper seal with your mask. You should clean any small nicks, scratches, and cuts with soap and water followed by first aid treatment.

Field Sanitation

13-35. Another way to stop the spread of disease is to keep the area clean. Use field sanitation facilities properly. Latrine facilities should include soap and water for hand washing. Avoid leaving such facilities open to help control the insects and rodents. This is also essential in preventing the spread of disease.

Physical Condition

13-36. Good physical conditioning requires that you keep your body well-rested, well-fed, and healthy. Get as much exercise and rest as the situation permits, and remember to eat properly. You may have to eat smaller portions, but at more frequent intervals. For you to be able to fight off germs, you must remain healthy.

Department of Defense Insect Repellent System

13-37. Proper implementation of the DOD insect repellent system will provide you protection from insects and ticks, which can serve as vectors, spreading biological agents.

Reinforcement Training

13-38. Training in an CBRN environment should be integrated into all areas of your unit training. Reinforcement training both individual and collective tasks should be performed to standard through continuous training, thereby instilling your individual confidence. Your life and the lives of your fellow Soldiers could depend on it.

Section III. RADIOLOGICAL WEAPONS

A radiological dispersal device (RDD) is a conventional bomb, not a yield producing nuclear device. An RDD disperses radioactive material to destroy, contaminate, and injure. An RDD can be almost any size.

TYPES

13-39. The types of radiological dispersal devices (RDDs) follow:

Passive

13-40. A passive RDD is unshielded radioactive material dispersed or placed manually at the target.

Chapter 13

EXPLOSIVE

13-41. The main employment of RDDs is explosives, which can cause serious injuries and property damage. An explosive RDD, often called "dirty bomb," is any system that uses the explosive force of detonation to disperse radioactive material. A dirty bomb uses dynamite or other explosives to scatter radioactive dust, smoke, or other materials in order to cause radioactive contamination.

13-42. A simple explosive RDD, commonly called a *pig*, has a lead-shielded container with a kilogram of explosive attached. A *pig* can easily fit into a backpack. The radioactive materials in an RDD probably produces too little exposure to cause immediate serious illness, except to those who are very close to the blast site. However, radioactive dust and smoke, when spread farther away, could endanger health if inhaled. Terrorist use of RDDs could cause health, environmental, political, social, and economic effects. It could also cause fear, and could cost much money and time to clean up.

ATMOSPHERIC

13-43. An atmospheric RDD is any device that converts radioactive materials into a form that is easily transported by air currents.

DETECTION

13-44. Because you cannot see, smell, feel, or taste radiation, you may not know whether you have been exposed. Low levels of radiation exposure--like those expected from a dirty bomb--cause no symptoms. Higher levels may produce symptoms such as nausea, vomiting, diarrhea, and swelling and redness of the skin.

DECONTAMINATION

13-45. Decontaminate and treat casualties the same as you would for exposure to nuclear radiation.

PROTECTION

13-46. Should you know when an incident occurs, take immediate steps to protect yourself:

OUTSIDE

- Cover your nose and mouth with a cloth to reduce the risk of breathing in radioactive dust or smoke.
- Don't touch objects thrown off by an explosion as they might be radioactive.
- Quickly go into a building where the walls and windows have not been broken. This area will shield you from radiation that might be outside.
- Once you are inside, take off your outer layer of clothing and seal it in a plastic bag if available. Put the cloth you used to cover your mouth in the bag as well. Removing outer clothes may get rid of up to 90% of radioactive dust.
- Shower or wash with soap and water. Be sure to wash your hair. Washing will remove any remaining dust.

INSIDE

- If the walls and windows of the building are not broken, stay in the building and do not leave.
- To keep radioactive dust or powder from getting inside, shut all windows and outside doors. Turn off fans and heating and air conditioning systems that bring in air from the outside. It is not necessary to put duct tape or plastic around doors or windows.

- If the walls and windows of the building are broken, go to an interior room and do not leave. If the building has been heavily damaged, quickly go into a building where the walls and windows have not been broken. If you must go outside, be sure to cover your nose and mouth with a cloth. Once you are inside, take off your outer layer of clothing and seal it in a plastic bag if available. Store the bag where others will not touch it.
- Shower or wash with soap and water, removing any remaining dust. Be sure to wash your hair.

IN A VEHICLE

- Close the windows and turn off the air conditioner, heater, and vents.
- Cover your nose and mouth with a cloth to avoid breathing radioactive dust or smoke.
- If you are close to a building, go there immediately and go inside quickly.
- If you cannot get to another building safely, pull over to the side of the road and stop in the safest place possible. If it is a hot or sunny day, try to stop under a bridge or in a shady spot.
- Turn off the engine.

Section IV. NUCLEAR WEAPONS

When a nuclear explosion occurs, blast radiation, and heat or thermal effects will occur. When a nuclear weapon detonates at low altitudes, a fireball results from the sudden release of immense quantities of energy. The initial temperature of the fireball ranges into millions of degrees, and the initial pressure ranges to millions of atmospheres. Characteristics of nuclear explosions and their effects on Soldiers, equipment, and supplies, and hasty measures for protection against nuclear attacks will be discussed in this section.

CHARACTERISTICS

13-47. Nuclear explosions are comprised of the following:

BLAST (INTENSE SHOCK WAVE)

13-48. Blast produces an intense shock wave and high winds, causing debris to fly. The force of a nuclear blast can collapse shelters and fighting positions.

THERMAL RADIATION (HEAT AND LIGHT)

13-49. Thermal radiation starts fires and causes burns. The bright flash at the time of the explosion can cause a temporary loss of vision or permanent eye damage if you look at the explosion, especially at night.

NUCLEAR RADIATION (RADIOACTIVE MATERIAL)

13-50. Nuclear radiation can cause casualties and delay movements. It may last for days and cover large areas of terrain. It occurs in two stages:

Initial Radiation

13-51. This type of radiation emits directly from the fireball in the first minute after the explosion. It travels at the speed of light along straight lines and has high penetrating power.

Residual Radiation

13-52. This type of radiation lingers after the first minute. It comes from the radioactive material originally in a nuclear weapon or from material, such as soil and equipment, made radioactive by the nuclear explosion.

ELECTROMAGNETIC PULSE

13-53. An EMP is a massive surge of electrical power. It is created the instant a nuclear detonation occurs, and it travels at the speed of light in all directions. It can damage solid state components of electrical equipment, such as radios, radar, computers, vehicles--and weapon systems. You can protect equipment by disconnecting it from its power source and placing it in or behind some type of shielding material, such as an armored vehicle or dirt wall, out of the line of sight from the burst. Without warning, there is no way for you to protect your equipment.

DETECTION

13-54. Radiation is the only direct nuclear effect that lingers after the explosion. As it cannot be detected by the senses, use radiac equipment to detect its presence (FM 3-11.3).

RADIAC SET AN/VDR 2

13-55. The AN/VDR 2 is used to perform ground radiological surveys in vehicles or in the dismounted mode by individual Soldiers as a handheld instrument (Figure 13-12). The set can also provide a quantitative measure of radiation to decontaminate personnel, equipment, and supplies. The set includes an audible and/or visual alarm that is compatible with vehicular nuclear, biological, and chemical protective systems in armored vehicles and also interfaces with vehicular power systems and intercoms.

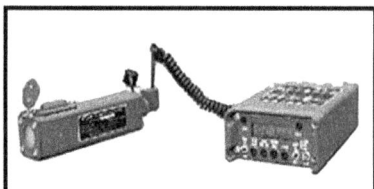

Figure 13-12. Radiac Set AN/VDR 2.

RADIAC SET AN/UDR 13

13-56. The AN/UDR 13 is a compact, handheld, or pocket carried, tactical device that can measure prompt gamma/neutron doses from a nuclear event, plus gamma dose and dose rate from nuclear fallout (Figure 13-13). A push-button pad enables mode selection, functional control, and the setting of audio and visual alarm thresholds for both dose rate and mission dose. A sleep mode with automatic wakeup lengthens battery life. The LCD provides data readout and warning and mode messages.

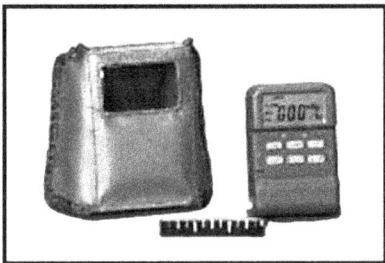

Figure 13-13. Radiac set AN/UDR 13.

13-57. If detection equipment is unavailable and you suspect that you are contaminated, decontaminate as required. Procedures for decontamination operations can be found in FM 3-11.5. Radiological contamination can usually be removed by brushing or scraping. When feasible, move out of the contaminated area.

13-58. If your unit must remain in the contaminated area, you should stay in a dug-in position with OHC. If you have time, brush or scoop away the top inch of soil from your fighting position to lower the amount of radiological contamination affecting you. When time does not permit constructing a well prepared OHC, use a poncho. Stay under cover. When the fallout is over, brush contamination off yourself and your equipment. Use water to flush away radiological contamination. However, control the runoff by using drainage ditches that flow into a sump. As soon as mission permits, wash yourself and your equipment. Remember, you have not destroyed the contamination; you have just moved it. The runoff will still be hazardous.

DECONTAMINATION

13-59. Blasts, and thermal and nuclear radiation causes nuclear casualties. Except for radiation casualties, treat nuclear casualties the same as conventional casualties. Wounds caused by blast are similar to other combat wounds. Thermal burns are treated as any other type of burn. The exposure of the human body to nuclear radiation causes damage to the cells in all parts of the body. This damage is the cause of "radiation sickness." The severity of this sickness depends on the radiation dose received, the length of exposure, and the condition of the body at the time. The early symptoms of radiation sickness will usually appear 1 to 6 hours after exposure. Those symptoms may include headache, nausea, vomiting, and diarrhea. Early symptoms may then be followed by a latent period in which the symptoms disappear. There is no first aid for you once you have been exposed to nuclear radiation. The only help is to get as comfortable as possible while undergoing the early symptoms. If the radiation dose was small, the symptoms, if any, will probably go away and not recur. If the symptoms recur after a latent period, you should go to an aid station. A blast can crush sealed or partly sealed objects like food cans, barrels, fuel tanks, and helicopters. Rubble from buildings being knocked down can bury supplies and equipment. Heat can ignite dry wood, fuel, tarpaulins, and other flammable material. Light can damage eyesight. Radiation can contaminate food and water.

PROTECTION

13-60. An attack occurring without warning is immediately noticeable. The first indication will be very intense light. Heat and initial radiation come with the light, and the blast follows within seconds. Nuclear attack indicators are unmistakable. The bright flash, enormous explosion, high winds, and mushroom shaped cloud clearly indicate a nuclear attack. An enemy attack would normally come without warning. Initial actions must, therefore, be automatic and instinctive. The best hasty protection against a nuclear attack is to take cover behind a hill or in a fighting position, culvert, or ditch. Time available to take protective action will be minimal. When in a fighting position, you can take additional precautions. The fighting position puts more earth between you and the potential source of radiation. You can curl up on one side, but the best position is on the back with knees drawn up to the chest. This position may seem vulnerable, but the arms and legs are more radiation resistant and will protect the head and trunk. However, if you're exposed while in the open when a detonation occurs, you should do the following:

- Drop face down immediately, with your feet facing the blast. This will lessen the possibility of heat and blast injuries to your head, face, and neck. A log, a large rock, or any depression in the earth's surface provides some protection.
- Close eyes.
- Protect exposed skin from heat by putting hands and arms under or near the body and keeping the helmet on.
- Remain facedown until the blast wave passes and debris stops falling.
- Remain calm, check for injury, check weapons and equipment for damage, and prepare to continue the mission.

Chapter 14
Mines, Demolitions, and Breaching Procedures

A unit may use mines during security, defensive, and offensive operations in order to reduce the enemy's mobility. In such operations, you must emplace the mines and, when required, retrieve them. In order to breach minefields and wire obstacles; there will be times when you have to physically detect and clear them. You must be proficient at correctly and safely handling demolition firing systems. This chapter will help you gain a true appreciation of the requirements and time it takes to perform an actual mine-warfare mission.

Section I. MINES

This section discusses antipersonnel and antitank mines. Some mines are "smart." That is, they contain RF receivers, which allow for remote or automatic self-destruction or self-deactivatation via a remote control unit (RCU), on demand, after a period of time, or at a particular time. The default self-destruct time, once the mines are dispensed, is four hours.

ANTIPERSONNEL MINES

14-1. Antipersonnel mines are designed specifically to reroute, block, or protect friendly obstacles. These mines are designed to kill or disable their victims, and are activated by command detonation.

**US NATIONAL POLICY
ON ANTIPERSONNEL LAND MINES**

On May 16, 1996, the President of the United States implemented a phased restriction and elimination of antipersonnel land mines. Implementation began with non-self-destructing mines, but will eventually include all types of antipersonnel mines. This policy applies to all Infantry units either engaged in or training for operations worldwide. The use of non-self-destructing antipersonnel land mines is restricted to specific areas:

- Within internationally recognized national borders.
- In established demilitarized zones such as to defend South Korea.

Mines approved for use must be emplaced in an area with clearly marked perimeters. They must be monitored by military personnel and protected by adequate means to ensure the exclusion of civilians.

US policy also forbids US forces from using standard or improvised explosive devices as booby traps.

Except for South Korea-based units, and for units deploying to South Korea for training exercises, this policy forbids training with and employing inert M14 and M16 mines. This applies to units' home stations as well as at Combat Training Centers, except in the context of countermine or mine removal training.

- *Training with live M14 mines is UNAUTHORIZED!*
- *Training with live M16 mines is authorized only for Soldiers on South Korean soil.*

Exceptions:

This policy does not apply to standard use of antivehicular mines. Nor does it apply to training and using the M18 Claymore mine *in the command-detonated mode*.

When authorized by the appropriate commander, units may still use self-destructing antipersonnel mines such as the ADAM.

Authorized units may continue to emplace mixed minefields containing self-destructing antipersonnel land mines and antivehicular land mines such as MOPMS or Volcano.

The terms *mine, antipersonnel obstacle, protective minefield,* and *minefield* do not refer to an obstacle that contains non-self-destructing antipersonnel land mines or booby traps.

Any references to antipersonnel mines and the employment of minefields should be considered in the context of this policy.

M18A1 ANTIPERSONNEL MINE (CLAYMORE)

14-2. The M18A1, also known as the Claymore mine, is a directional, fragmentating (one-time use) antipersonnel mine (Figure 14-1, Figure 14-2, and Figure 14-3 [page 14-4]). The Claymore weighs 1.6 kilograms (3.5 pounds), 0.68 kilograms (1.5 pounds) of which is C4 (explosive) and steel sphere projectiles. One Claymore and its accessories are carried in the M7 bandoleer (Figure 14-2). When detonated, the Claymore projects steel fragments over a 60-degree, fan-shaped pattern about 6 feet (1.8 meters) high and 164 feet (50 meters) wide, at a range of 50 meters. This pattern of distribution is

Mines, Demolitions, and Breaching Procedures

very effective up to 50 meters, moderately effective to 100 meters, and still dangerous out to 250 meters. It has the following features:
- A fixed plastic sight.
- Folding, adjustable legs.
- Two detonator wells.
- Olive-drab plastic case.

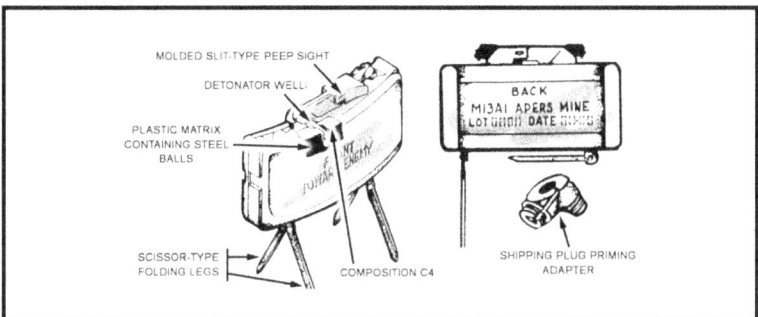

Figure 14-1. M18A1 antipersonnel mine.

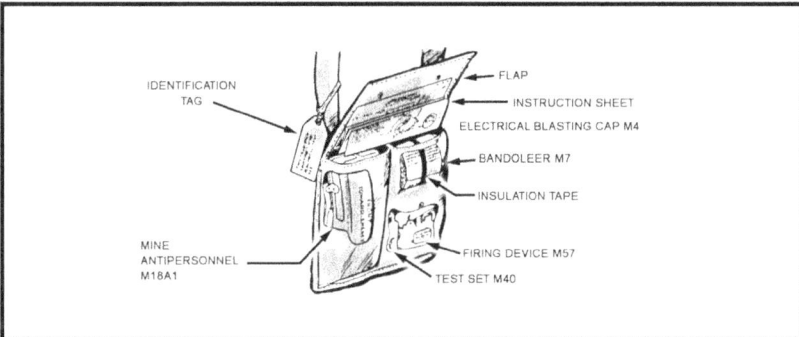

Figure 14-2. M7 bandoleer.

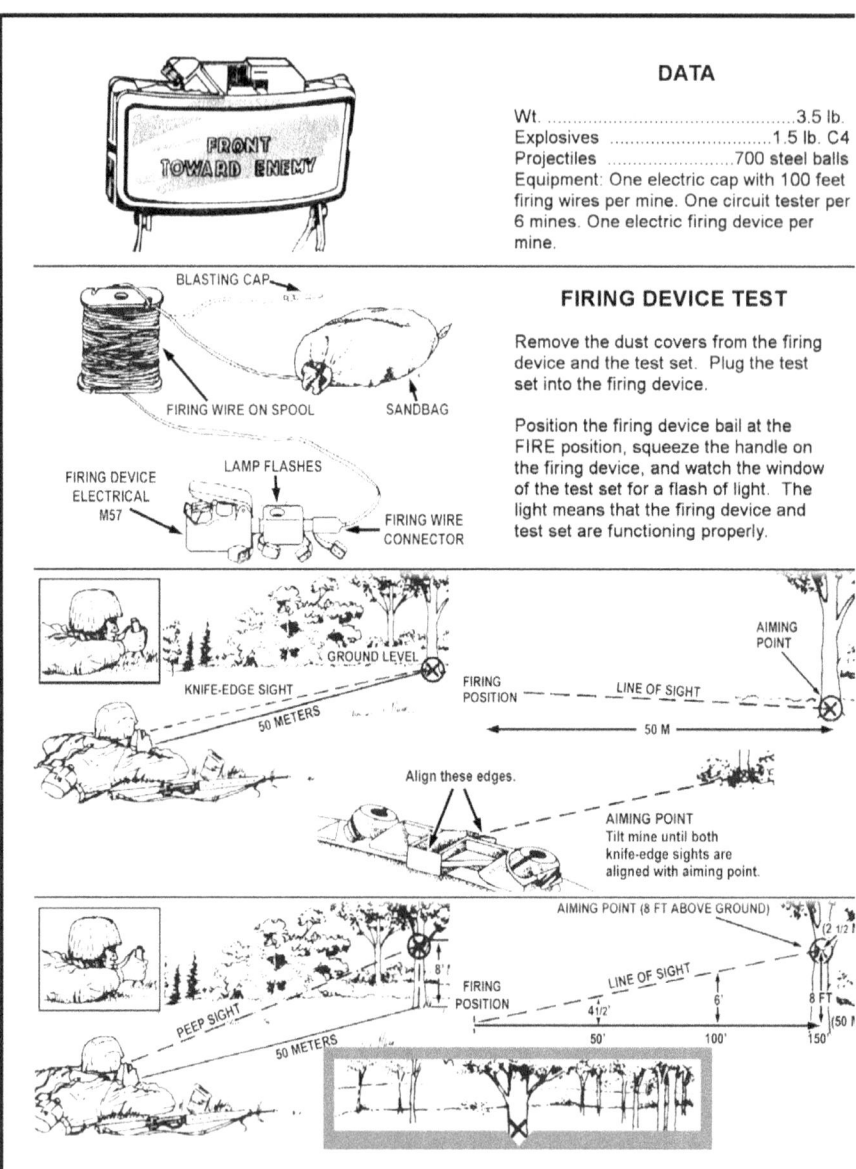

Figure 14-3. M18A1 antipersonnel mine data.

Mines, Demolitions, and Breaching Procedures

Employment

14-3. The Claymore is mainly a defensive weapon used to support other weapons in a unit's final protective fires. The Claymore may also be employed in some phases of offensive operations. Complete instructions for installing, arming, testing, and firing the Claymore are attached to the flap of the bandoleer. If possible, read the directions before employing the mine.

Emplacement for Command Detonation

14-4. Inventory the M7 bandoleer, ensuring all components of the M18A1 Claymore mine are present and in serviceable condition. Components consist of an M18A1 Claymore mine, M57 firing device, M40 test set, and a firing wire with blasting cap.

- Conduct a circuit test, with the blasting cap secured under a sandbag.
- Install the M18A1 Claymore mine.
- Aim the mine.
- Arm the mine.
- Recheck the aim.
- Recheck the circuit.
- Fire the M18A1 Claymore mine.

Recovery

- Check the firing device safety bail to ensure it is on SAFE.
- Disconnect the firing device from the wire.
- Replace the shorting plug dustcover on the firing wire connector.
- Replace the dustcover on the firing device connector.
- Keep possession of the M57 firing device.
- Untie the firing wire from the firing site stake.
- Move to the mine.
- Remove the shipping plug before priming.
- Separate the blasting cap from the mine.
- Reverse the shipping plug.
- Screw the shipping plug end into the detonator well.
- Remove the firing wire from the site stake.
- Place the blasting cap into the end of the wire connector.
- Roll the firing wire on the wire connector.
- Lift the mine from its emplacement.
- Secure the folding legs.
- Repack the mine and all the accessories in the M7 bandoleer.

M-131 MODULAR PACK MINE SYSTEM

14-5. The MOPMS is a man-portable antitank and antipersonnel mine system (Figure 14-4). The M-131 module weighs about 165 pounds (75 kilograms), and contains a mix of 17 M78 antiarmor and 4 M77 antipersonnel mines. The MOPMS module may be initiated by hardwire or radio control. The hardwire capability uses wire and electrical firing devices. The M-71 handheld radio control unit (RCU) allows one Soldier to control as many as 15 groups of MOPMS modules from a remote location.

Chapter 14

Figure 14-4. M-131 Modular Pack Mine System (MOPMS).

14-6. This mine deploys four trip wires upon ejection. These wires trip a fragmenting kill mechanism (Figure 14-5).

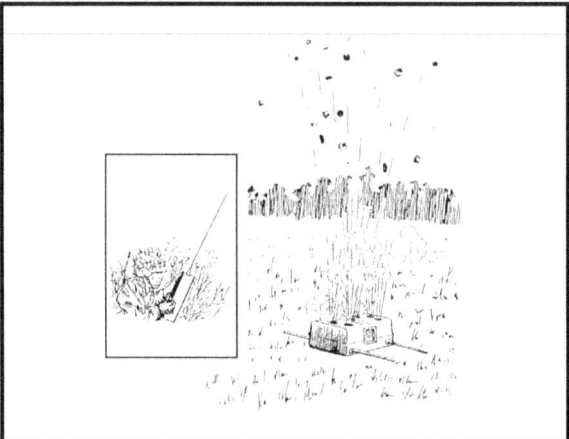

Figure 14-5. M-131 MOPMS deployed.

14-7. Each dispenser contains seven tubes; three mines are located in each tube. When dispensed, an explosive propelling charge at the bottom of each tube expels mines through the container roof. Mines are propelled 115 feet (35 meters) from the container in a 180-degree semicircle (Figure 14-6). The resulting density is 0.01 mine per square meter. The safety zone around one container is 180 feet (55 meters) to the front and sides and 66 feet (20 meters) to the rear.

Mines, Demolitions, and Breaching Procedures

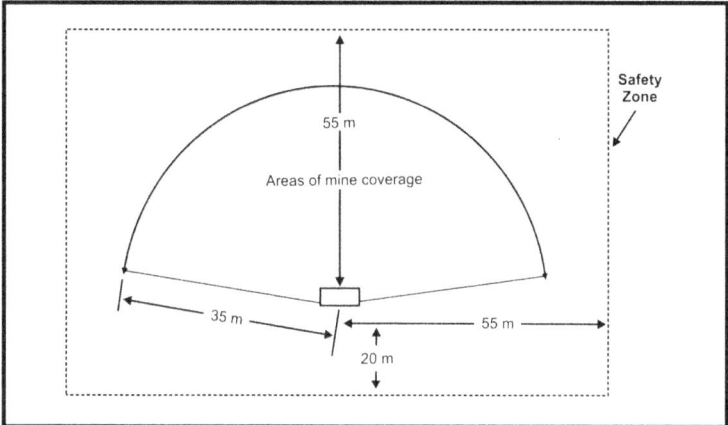

Figure 14-6. MOPMS emplacement and safety zone.

14-8. You can disarm and recover the container for later use, if mines are not dispensed. The RCU can recycle the 4-hour self-destruct time of the mines three times, for a total duration of 16 hours (4 hours after initial launch and three 4-hour recycles). This feature makes it possible to keep the minefield in place for longer periods, if necessary. The RC can also self-destruct mines on command. However, once the mines are dispensed they cannot be recovered or reused.

M21 ANTITANK MINE

14-9. Antitank mines are pressure activated, but are typically designed so that the force of a footstep will not detonate them. Most antiarmor mines require an applied pressure of 348 to 745 pounds (158 to 338 kilograms) in order to detonate. Most tanks and other military vehicles apply that kind of pressure. Most antiarmor mines go off on contact, but some are designed to count a preset number of pressures before going off. By delaying detonation, a large number of enemy vehicles and troops might travel deep within the minefield before knowing the area is dangerous. The US policy for the use of non-self-destructing antitank mines applies until 2010, when these mines must be replaced by models that either self-destruct or at least self-deactivate. The M21 is a circular, steel bodied, antiarmor mine designed to damage or destroy vehicles by a penetrating effect (Figure 14-7). The bottom of the mine is crimped to the upper mine body. An adjustable cloth carrying handle is attached to the side of the mine body and a large filler plug is positioned between the handle connection points. A booster well is centered on the bottom. The mine has a small diameter fuse cavity and a stamped radial pattern centered on top. The M21 is 9 inches (23 centimeters) in diameter and 4 1/2 inches (11 centimeters) high. It weighs a total of 17 pounds (8 kilograms) with 11 pounds (5 kilograms) of high explosives. The M607 fuse protrudes from the top. This mechanical fuse can be used with or without a tilt rod. Without the tilt rod, it works like a normal pressure fuse.

Chapter 14

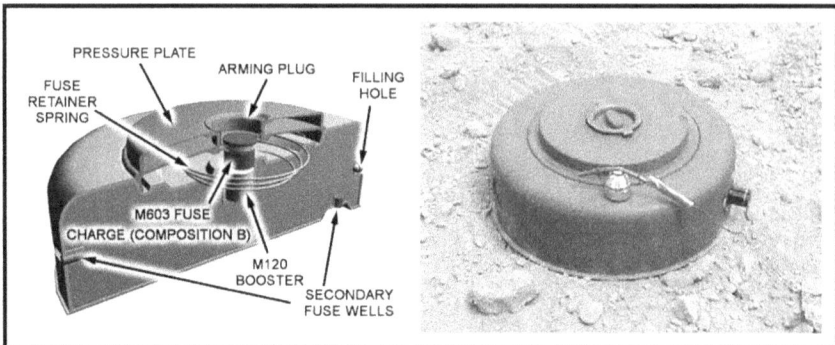

Figure 14-7. M21 antitank mine and components.

14-10. The M21 is activated by 1.8 kilograms (4 pounds) of pressure against a 21-inch (53-centimeter) long extension rod or, without the rod, by 290 pounds (132 kilograms) of vertical pressure on top of the M607 fuse. Once the fuse is triggered, it releases a firing pin, which is driven into the M46 detonator, which in turn sets off a small, black powder charge. This charge blows off the top of the mine, exposing a convex steel plate. It also drives another firing pin into an M42 primer, which in turn fires the main charge. The main charge blows the body apart and blasts the steel plate upwards through the belly armor of the tank. Unlike most antiarmor mines, this one can actually kill a tank, not just disable it. It uses a Miznay-Schardin plate as a directed-energy warhead; a kill mechanism for belly-kill and track-breaking capability. The M21 produces a kill against heavy tanks, unless the mine is activated under the track.

Section II. DEMOLITION FIRING SYSTEMS

The modernized demolition initiator (MDI) is a suite of initiating components used to activate all standard military demolitions and explosives (Figure 14-8). The MDI consists of nonelectric blasting cap assemblies (M11/12/13/14/15/16/18) each with an integral time-delay initiator: a time fuse or shock tube; and a "J" hook for attachment to a detonating cord. These MDIs will eventually replace all electric and nonelectric firing systems for conventional forces, while maintaining compatibility with existing Army systems.

Note: Information on the preparation and placement of demolition charges for electric and nonelectric firing systems separate from MDI is in FM 3-34.214.

BOOSTER ASSEMBLIES

14-11. The MDI includes a pair of booster assemblies (M151/M152), consisting of a detonator (det) and a length of low-strength detonation cord, were added to MDI. Since they contain no sensitive initiating element, they can be used to safely initiate underground charges. The MDI's blasting cap assemblies consist of five high-strength blasting caps and two high strength booster assemblies. These cap assemblies can be used to prime standard military explosives, or to initiate the shock tube or detonation cord of other MDI components. MDI also contains an igniter (M81) that can activate either time fuse or shock tube. With MDI, you can successfully complete demolition missions in a safe, quick, and easy manner. It is also flexible enough that any unit conducting demolition activities can use it. Timing can be set to fire immediately or up to a 20-minute delay. It is nonelectrical, so it is impervious to EMP.

Mines, Demolitions, and Breaching Procedures

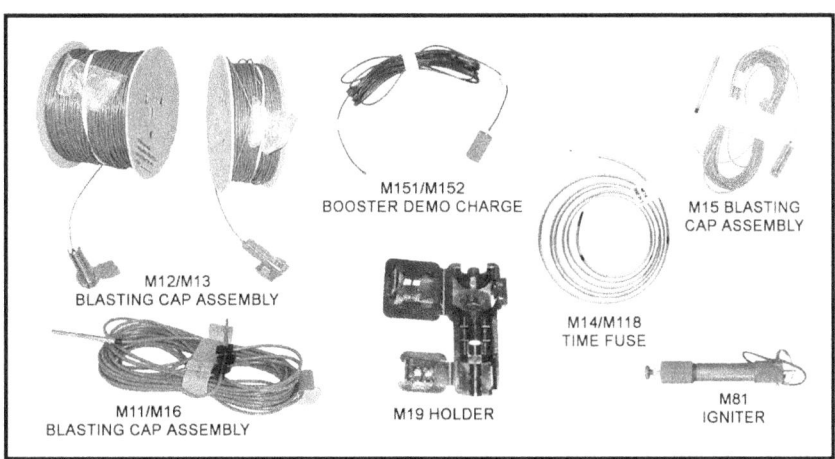

Figure 14-8. MDI components.

PRIMING OF EXPLOSIVES

14-12. The two methods of priming explosive charges are nonelectric (MDI) and detonating cord. MDI priming is safer and more reliable than the current nonelectric cap priming methods. However, detonating cord is the most preferred method of priming charges since it involves fewer blasting caps and makes priming and misfire investigation safer.

14-13. MDI blasting caps are factory-crimped to precut lengths of shock tube or time-blasting fuse. Because the caps are sealed, they are resistant to moisture and will not misfire in damp conditions. Splicing compromises the integrity of the system, and moisture will greatly reduce reliability. Also, the human factor in incorrect crimping is removed, making MDI blasting caps extremely reliable.

PRIMING OF NONELECTRIC MDI

14-14. Use only high-strength MDI blasting caps (M11, M16, M14, M15, or M18) to prime explosive charges. M12 and M13 relay-type blasting caps have too little power to reliably detonate most explosives. Use them only as transmission lines in firing systems. You can use MDI blasting caps with priming adapters, or you can insert them directly into the explosive charge, and then secure them with black electrical tape. If you use priming adapters, place them on M11 blasting caps as follows:

Priming Plastic Explosives with Nonelectric MDI--(See Figure 14-9.)

1. Use M2 crimpers or other nonsparking tools to make a hole in one end or on the side (at the midpoint) of the M112 (C4) demolition block. The hole should be large enough to hold an M11, M16, M14, M18, or M15 blasting cap.
2. Insert an M11, M16, M14, M18, or M15 blasting cap into the hole produced by the M2 crimpers.

Chapter 14

> **WARNING**
>
> If the blasting cap does not fit, *do not force it!*
> Instead, *make the hole larger.*

3. Anchor the blasting cap in the demolition block by gently squeezing the C4 plastic explosive around the blasting cap.
4. Use tape to secure the cap in the charge M112.

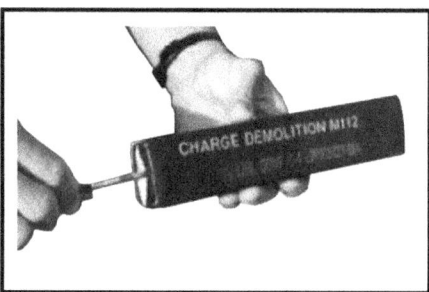

Figure 14-9. Priming of C4 demolition blocks with MDI.

PRIMING OF DETONATING CORD

1. Form either a uli knot, double overhand knot, or triple roll knot as shown in Figure 14-10.

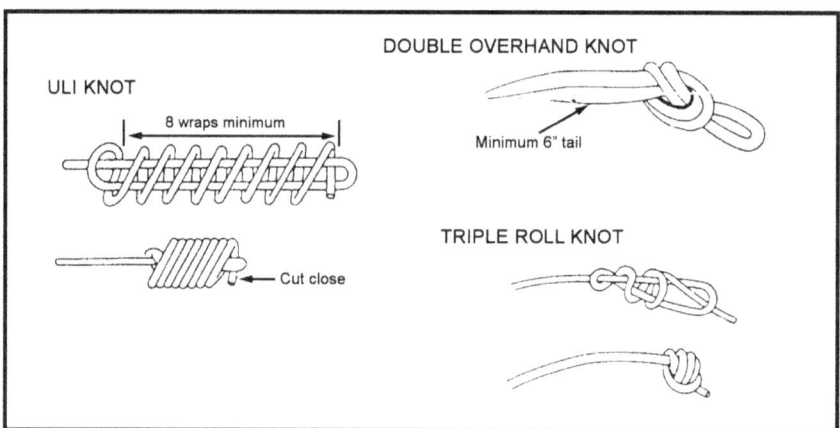

Figure 14-10. Priming of C4 demolition blocks with detonating cord.

Mines, Demolitions, and Breaching Procedures

2. Cut an L-shaped portion of the explosive, still leaving it connected to the explosive. Ensure the space is large enough to insert the knot you formed (Figure 14-11).
3. Place the knot in the L-shaped cut.
4. Push the explosive from the L-shaped cut over the knot. Ensure there is at least 1 centimeter (1/2 inches) of explosive on all sides of the knot.
5. Strengthen the primed area by wrapping it with tape.

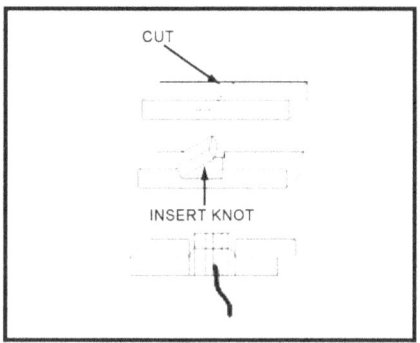
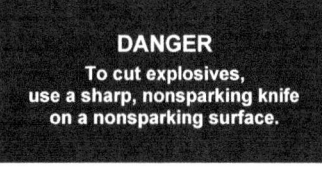

Figure 14-11.
Priming of C4 with L-shaped charge.

Note: Do not prime plastic explosives by wrapping them with detonating cord, since wraps will not properly detonate the explosive charge.

CONSTRUCTION OF NONELECTRIC INITIATING ASSEMBLY WITH MDI

1. Turn the end cap of the M81 fuse igniter a half-turn counterclockwise, and remove both the shipping plug and the shock tube adapter from the igniter (Figure 14-12).

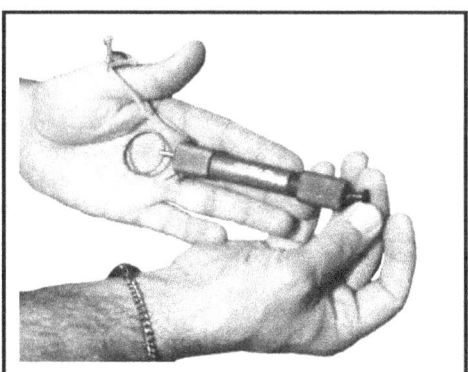

Figure 14-12. Preparation of M81 fuse igniter.

Chapter 14

2. Cut off the sealed end of the M14 time-delay fuse (Figure 14-13), and insert it into the end cap of the M81. Tighten (finger-tight) by turning the end cap clockwise.

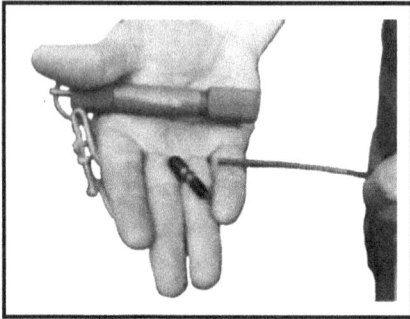

Figure 14-13. M81 fuse igniter with the M14 time fuse delay.

3. Attach the blasting-cap end of the M14 time-delay fuse to the existing detonating-cord ring/line main using either an M9 holder or adhesive tape. If using tape, ensure it is at least 6 inches from the end of the detonating cord.
   a. Attach the M14 blasting cap using the M9 holder (the preferred method).
   b. Open both hinged flaps of the M9 holder.
   c. Insert the blasting cap into the M9 holder and close the small hinged flap.
   d. Form a bight 6 inches from the end of the detonating cord, lay it in the M9 holder, and close the hinged flap.
   e. Secure the detonating cord into the M9 holder (Figure 14-14). Secure the door with adhesive tape.

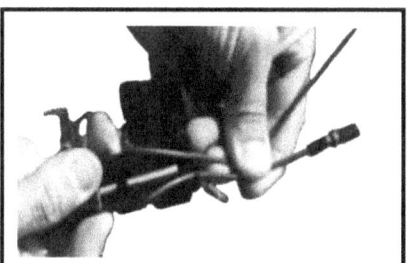

Figure 14-14. M81 fuse igniter with the M9 holder.

Note: Do not loop more than two shock tubes in the M9 holder.

4. Construct a nonelectric initiating assembly using the M11 branch line and the M12 transmission line.
   a. Place the M11 branch line's blasting cap under a sandbag near the detonating-cord firing system.

Mines, Demolitions, and Breaching Procedures

b. Attach the M11 branch line to the M12 transmission line by forming a bight at the end of the M11, laying it in the attached M9 holder on the M12, and closing the hinged flap. Tape and secure the M11 into place. Place the M9 holder, along with the M12, under the same sandbag as the M11 blasting cap.
c. Retrieve the M11 blasting cap from under the sandbag. Attach it to the detonating-cord firing system using an M9 holder as described above using either the M14 or adhesive tape. Ensure the tape is at least 6 inches from the end of the detonating cord.
d. Secure the transmission line to a nearby anchor point and run the M12 transmission line back to the initiating point.

5. Cut the sealed end of the M12 transmission line at the initiating point, and attach an M81 fuse igniter as described above for the M14 time-delay fuse (Figure 14-15).

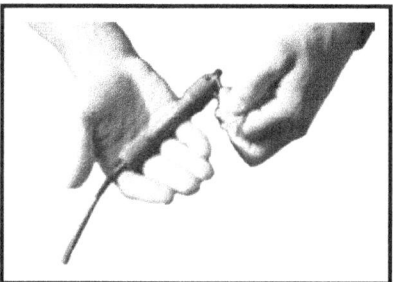

Figure 14-15. M81 fuse igniter with the M14 time fuse delay.

Note: MDI systems come with short clear plastic tubes used for repair. The shock tube repair procedure is outlined in TM 9-1375-213-12, Army Demolition.

6. Firing procedure for a nonelectric initiating assembly with MDI.
   a. Squeeze the spread legs of the safety cotter pin together.
   b. Use the safety pin's cord to remove the safety cotter pin from the igniter's body.
   c. Grasp the igniter body firmly with one hand, with the pull ring fully accessible to the other hand. To actuate, sharply pull the igniter's pull ring. The igniter can burn at extremely high temperatures.
   d. Ensuring that smoke is coming from the fuse (or out of the vent hole in the igniter), remove the igniter and withdraw to a safe distance or to appropriate cover.

WARNING

When using MDI in extreme cold temperatures and/or high altitudes, dual prime and dual initiate the charges.

MISFIRES

14-15. The most common cause of a misfire in a shock-tube firing system is the initiating element, usually an M81 igniter. However, the most common failure with the M81 is primer failure to fire. To

Chapter 14

correct this, recock the M81 by pushing in on the pull rod to reengage the firing pin and then actuate the igniter again. If, after two retries, the M81 does not result in firing, cut the shock tube, replace the igniter with a new one, and repeat the firing procedure.

14-16. Another misfire mode with the M81 is the primer fires but blows the shock tube out of its securing mechanism without it firing. (This is usually due to the shock tube not being properly inserted and secured in the igniter.) To correct this problem, cut about 91 centimeters (3 feet) from the end of the shock tube, replace it with a new igniter, and repeat the firing procedure.

Note: Your supervisor needs to be involved to identify and correct any additional misfire problems. The correct procedure to implement for all possible misfires is in FM 3-34.214.

Section III. OBSTACLES

This section discusses how to breach and cross a minefield or wire obstacle.

BREACH AND CROSS A MINEFIELD

14-17. In combat, enemy units use obstacles to stop slow or channel their opponent's movement. Because of that, you may have to bypass or breach (make a gap through) those obstacles in order to continue your mission. There are many ways to breach a minefield. One way is to probe for and mark mines to clear a footpath through the minefield.

PROBING FOR MINES

1. Leave your rifle and LCE with another Soldier in the team.
2. Leave on your Kevlar helmet and vest to protect you from possible blasts.
3. Get a wooden stick about 30 centimeters (12 inches) long for a probe and sharpen one of the ends. *Do not use a metal probe.*

DANGER
Do *not* use a metal probe.

4. Place the unsharpened end of the probe in the palm of one hand with your fingers extended and your thumb holding the probe.
5. Probe every 5 centimeters (2 inches) across a 1-meter front. Push the probe gently into the ground at an angle less than 45 degrees (Figure 14-16).
6. Kneel (or lie down) and feel upward and forward with your free hand to find tripwires and pressure prongs before starting to probe.
7. Put just enough pressure on the probe to sink it slowly into the ground. If the probe does not go into the ground, pick or chip the dirt away with the probe and remove it by hand.

Mines, Demolitions, and Breaching Procedures

Figure 14-16. Mine probe.

8. Stop probing when the probe hits a solid object.
9. Remove enough dirt from around the object to find out what it is.
10. Clear a lane in depth of 10-meter (33 feet) intervals and ensure the lane overlaps (Figure 14-17).

Figure 14-17. Lanes.

Chapter 14

MARKING THE MINE

14-18. Remove enough dirt from around it to see what type of mine it is.

1. Mark it and report its exact location to your leader. There are several ways to mark a mine. How it is marked is not as important as having others understand the marking. A common way to mark a mine is to tie a piece of paper, cloth, or engineer tape to a stake and put the stake in the ground by the mine (Figure 14-18).

Figure 14-18. Knot toward mine.

CROSSING THE MINEFIELD

14-19. Once a footpath has been probed and the mines marked, a security team should cross the minefield to secure the far side (Figure 14-19). After the far side is secure, the rest of the unit should cross.

Figure 14-19. Marked mines.

Mines, Demolitions, and Breaching Procedures

BREACH AND CROSS A WIRE OBSTACLE

14-20. Breaching a wire obstacle may require stealth; for example, when done by a patrol. It may not require stealth during an attack. Breaches requiring stealth are normally done with wire cutters. Other breaches are normally done with bangalore torpedoes and breach kits. This paragraph discusses how to probe for and mark mines, as well as cross minefields:

CUTTING THE WIRE

14-21. This paragraph discusses how to cut and cross wire.

To cut through a wire obstacle with stealth--

1. Cut only the lower strands and leave the top strand in place; this decreases the likelihood that the enemy will discover the gap.
2. Cut the wire near a picket. To reduce the noise of a cut, have another Soldier wrap cloth around the wire and hold the wire with both hands. Cut part of the way through the wire between the other Soldier's hands and have him bend the wire back and forth until it breaks. If you are alone, wrap cloth around the wire near a picket, partially cut the wire, and then bend and break the wire.

To breach an obstacle made of concertina:

1. Cut the wire and stake it back to keep the breach open.
2. Stake the wire back far enough to allow room for Soldiers to move through the obstacle.

USING A BANGALORE TORPEDO

14-22. A bangalore torpedo comes in a kit that has ten torpedo sections, ten connecting sleeves, and one nose sleeve (Figure 14-20). Use only the number of torpedo sections and connecting sleeves needed.

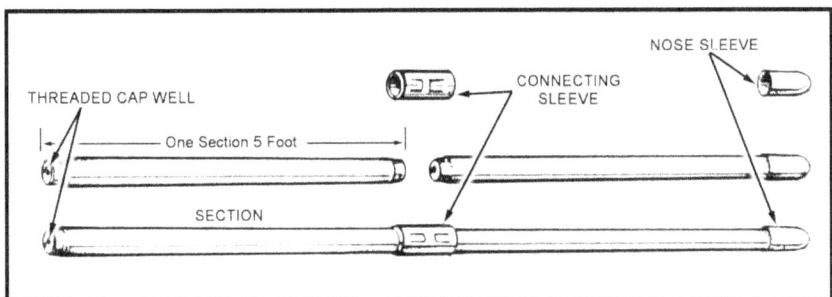

Figure 14-20. Bangalore torpedo.

14-23. All torpedo sections have a threaded cap well at each end so they may be assembled in any order. Use the connecting sleeves to connect the torpedo sections together. To prevent early detonation of the entire bangalore torpedo, should you actually hit a mine while pushing the bangalore through the obstacle, attach an improvised (wooden) torpedo section to its end. That section can be made out of any wooden pole or stick equal to the size of a real torpedo section. Attach the nose sleeve to the end of the wooden section.

14-24. After the bangalore torpedo has been assembled and pushed through the obstacle, prime it with either detonation cord or with the MDI nonelectric firing system (Figure 14-21); only the MDI method will be described. The bangalore torpedo is primed using an M11, M16, M14, or M18 blasting cap. Insert the blasting cap into the cap well in the end section of the charge and secure it with a priming adapter. If a priming adapter is unavailable, use tape to hold the blasting cap firmly in place.

Chapter 14

Figure 14-21. Priming of bangalore torpedo with MDI.

14-25. Before the bangalore torpedo is fired, ensure you seek available cover (at least 35 meters away) from the safety danger zone. You will use wire cutters to cut away any wire not cut by the explosion.

USING A MK7 ANTIPERSONNEL OBSTACLE-BREACHING SYSTEM

14-26. The APOBS is an explosive line charge system that allows safe breaching through complex antipersonnel obstacles (Figure 14-22). The APOBS is used to conduct deliberate or hasty breaches through enemy antipersonnel minefields and multistrand wire obstacles. A lightweight 125 pound (57 kilogram) system with delay and command firing modes, it can be carried by two Soldiers with backpacks and can be deployed within 30 to 120 seconds.

14-27. Once set in place, the APOBS rocket is fired from a 35-meter standoff position, sending the line charge with fragmentation grenades over the minefield and/or wire obstacle. The grenades neutralize or clear the mines and sever the wire, effectively clearing a footpath 2 to 3 feet (0.6 to 1 meter) wide by 148 feet (50 yards or 45 meters) long.

14-28. The APOBS has significant advantages over the bangalore torpedo, which weighs 145 kilograms (320 pounds) more, takes significantly longer to set up, and cannot be deployed from a standoff position. It also reduces the number of Soldiers required to carry and employ the system from 12 Soldiers to two.

Figure 14-22. MK7 Antipersonnel Obstacle-Breaching System (APOBS).

Chapter 15

Unexploded Ordnance and Improvised Explosive Devices

Unexploded ordnance (UXO) and IEDs pose deadly and pervasive threats to Soldiers and civilians in operational areas all over the world. Soldiers at all levels must know about these hazards, as well as how to identify, avoid, and react to them properly. Use the information in this chapter to learn about the UXO and IED hazards you could face and the procedures you can use to protect yourself.

Section I. UNEXPLODED ORDNANCE

Being able to recognize a UXO is the first and most important step in reacting to a UXO hazard. There is a multitude of ordnance used throughout the world, and it comes in all shapes and sizes. This chapter explains and shows some of the general identifying features of the different types of ordnance, both foreign and US. To learn more about UXOs. In this chapter, ordnance is divided into four main types: dropped, projected, thrown, and placed.

DROPPED ORDNANCE

15-1. Regardless of its type or purpose, dropped ordnance is dispensed or dropped from an aircraft. Dropped ordnance is divided into three subgroups: bombs; dispensers, which contain submunitions; and submunitions.

BOMBS

15-2. Bombs can be general purpose or chemical-agent filled.

General Purpose Bombs

15-3. General purpose bombs come in many shapes and sizes depending on the country that made them and how they are to be used. Most of these bombs are built the same and consist of a metal container, a fuse, and stabilizing device. The metal container (called the bomb body) holds an explosive or chemical filler. The body may be in one or multiple pieces.

Chemical-Agent Filled Bombs

15-4. Chemical-agent filled bombs are built the same as general purpose bombs. They have a chemical filler in place of an explosive filler. Color codes and markings may be used to identify chemical bombs. For example, the US and NATO color code for chemical munitions is a gray background with a dark green band. The former Soviet Union uses a combination of green, red, and blue markings to the nose and tail sections to indicate chemical agents. Soviet bombs all have a gray background.

Fuses

15-5. Fuses used to initiate bombs are either mechanical or electrical, and are generally placed in the bomb's nose or tail section, internally or externally. They may be hidden, as when covered by a fin assembly. As shipped, fuses are in a safe (unarmed) condition and function only once armed.

- *Mechanical*--Mechanical fusing, whether in the nose or in the tail, is generally armed by some type of arming vane. The arming vane assembly operates like a propeller to line up all of the fuse parts so the fuse will become armed.
- *Electrical*--Electrical fuses have an electric charging assembly in place of an arming vane. They are armed by using power from the aircraft. Just before the pilot releases the bomb, the aircraft supplies

the required electrical charge to the bomb's fuse. Action of the fuse may be impact, proximity, or delay. Impact fuses function when they hit the target. Proximity fuses function when bombs reach a predetermined height above the target.

Delay--Delay fuses contain an element that delays explosion for a fixed time after impact. To be safe, personnel should consider that all bombs have the most dangerous kind of fusing, proximity or delay. Approaching a proximity or delay-fused bomb causes unnecessary risk to personnel and equipment. Although it should function before it hits the target, proximity fusing may not always do so. Once the bomb hits the ground, the proximity fuse can still function. It can sense a change in the area around the bomb and detonate. Delay fusing can be mechanical, electrical, or chemical. Mechanical and electrical-delay fuses are nothing more than clockwork mechanisms. The chemical-delay fuse uses a chemical compound inside the fuse to cause a chemical reaction with the firing system. Delay fusing times can range from minutes to days.

DISPENSERS

15-6. Dispensers may be classified as another type of dropped ordnance. Like bombs, they are carried by aircraft. Their payload, however, is smaller ordnance called submunitions. Dispensers come in a variety of shapes and sizes depending on the payload inside. Some dispensers are reusable, and some are one-time-use items.

SUBMUNITIONS

15-7. Submunitions are classified as bomblets, grenades, or mines. They are small explosive-filled or chemical-filled items designed for saturation coverage of a large area. Each of these delivery systems disperses its payload of submunitions while still in flight, and the submunitions drop over the target. On the battlefield, submunitions are widely used in both offensive and defensive missions. Submunitions are used to destroy an enemy in place (impact) or to slow or prevent enemy movement away from or through an area (area denial). Impact submunitions go off when they hit the ground.

ANTIPERSONNEL BALL-TYPE SUBMUNITIONS

15-8. Area-denial submunitions, including FASCAM, have a limited active life and self-destruct after their active life has expired. The ball-type submunitions shown in Figure 15-1 are antipersonnel. They are very small and are delivered on known concentrations of enemy personnel.

Note: Never approach a dispenser or any part of a dispenser you find on the battlefield. The payload of submunitions always scatters in the area where the dispenser hit the ground.

Unexploded Ordnance and Improvised Explosive Devices

Figure 15-1. Antipersonnel, ball-type submunitions.

Area-Denial Submunitions--The submunition shown in Figure 15-2 is scattered across an area and, like a land mine, it will not blow up until pressure is put on it. They are area-denial, antipersonnel submunitions (FASCAM). These submunitions are delivered into areas for use as mines. When they hit the ground, trip wires kick out up to 20 feet from the mine. All area-denial submunitions use anti-disturbance fusing with self-destruct fusing as a backup. The self-destruct time can vary from a couple of hours to as long as several days.

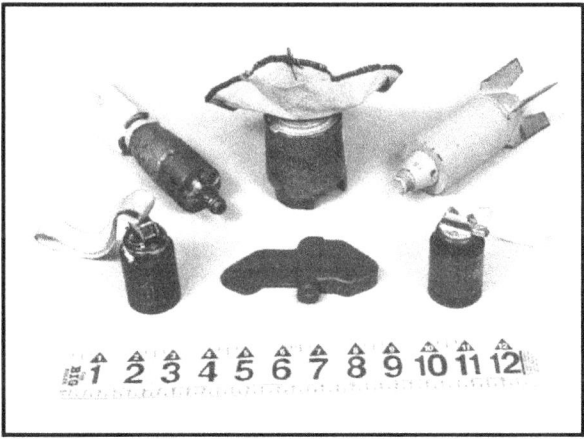

Figure 15-2. Area-denial submunitions (conventional).

Antipersonnel and Anti-Materiel (AMAT) Submunitions--The DP submunition shown in Figure 15-3 has a shaped charge for penetrating hard targets but is also used against personnel. These submunitions are delivered by artillery or rockets. The arming ribbon serves two purposes, as it arms the fuse as the submunition comes down and also stabilizes the submunition so that it hits the target straight on.

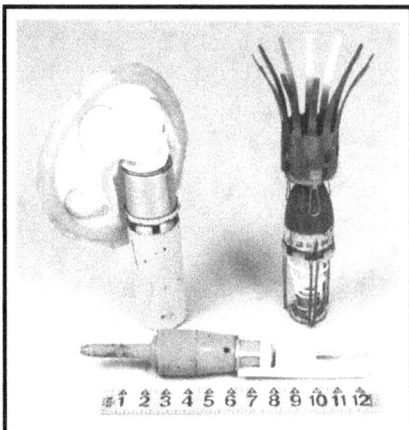

Figure 15-3. Antipersonnel/AMAT submunitions (conventional).

Unexploded Ordnance and Improvised Explosive Devices

AMAT and Antitank Submunitions--The AMAT or antitank submunitions shown in Figure 15-4 are designed to destroy hard targets such as vehicles and equipment. They are dispersed from an aircraft-dropped dispenser and function when they hit a target or the ground. Drogue parachutes stabilize these submunitions in flight so they hit their targets straight on. Others have a fin assembly that stabilizes the submunition instead of the drogue parachute.

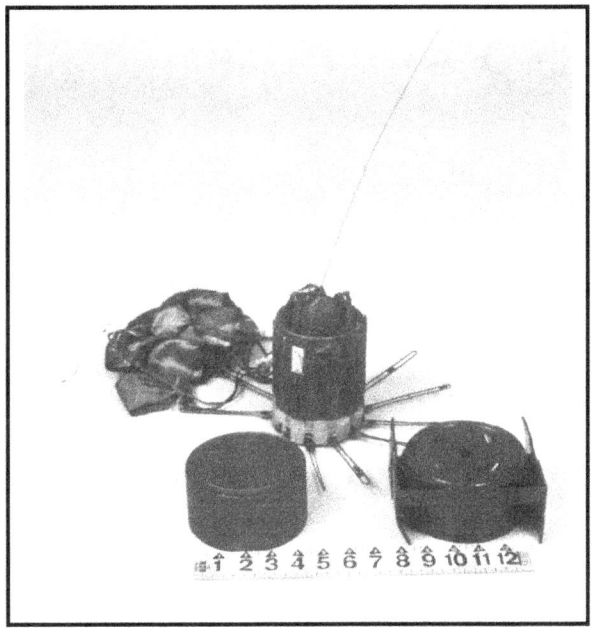

Figure 15-4. AMAT/antitank submunitions (conventional).

Antitank Area-Denial Submunitions--Antitank area-denial submunitions can be delivered by aircraft, artillery, and even some engineer vehicles. These FASCAMs have magnetic fusing and function when they receive a signal from metallic objects. These submunitions, similar to the antipersonnel area-denial submunitions, also have anti-disturbance and self-destruct fusing. Antitank and antipersonnel area-denial mines are usually found deployed together.

Chapter 15

PROJECTED ORDNANCE

15-9. All projected ordnance is fired from some type of launcher or gun tube. Projected ordnance falls into the following five subgroups:
- Projectiles.
- Mortars.
- Rockets.
- Guided missiles.
- Rifle grenades.

PROJECTILES

15-10. Projectiles range from 20 millimeters to 16 inches in diameter and from 2 inches to 4 feet in length. They can be filled with explosives, chemicals (to include riot-control agents such as CS), white phosphorus (WP), illumination flares, or submunitions. Projectile bodies can be one piece of metal or multiple sections fastened together.

15-11. Projectiles, like bombs, can have impact or proximity fusing. They can also be fused with time-delay fusing that functions at a preset time after firing. For safety reasons, all projectiles should be considered as having proximity fusing. Getting too close to proximity fusing will cause the fuse to function, and the projectile will blow up. Depending on the type of filler and design of the projectile, the fuse can be in the nose or base.

15-12. There are two ways projectiles are stabilized— by spin or fin. Spin-stabilized projectiles use rotating bands near the rear section to stabilize the projectile. Riding along the internal lands and grooves of the gun tube, these bands create a stabilizing spin as the projectile is fired. Fin-stabilized projectiles may have either fixed fins or folding fins. Folding fins unfold after the projectile leaves the gun tube to stabilize the projectile.

MORTARS

15-13. Mortars range from 45 millimeters to 280 millimeters in diameter. Like projectiles, mortar shells can be filled with explosives, toxic chemicals, WP, or illumination flares. Mortars generally have thinner metal bodies than projectiles but use the same kind of fusing. Like projectiles, mortars are stabilized in flight by fin or spin. Most mortars are fin stabilized.

ROCKETS

15-14. A rocket may be defined as a self-propelled projectile. Unlike guided missiles, rockets cannot be controlled in flight. Rockets range in diameter from 37 millimeters to over 380 millimeters. They can range in length from 1 foot to over 9 feet. There is no standard shape or size for rockets. All rockets consist of a warhead section, motor section, and fuse. They are stabilized in flight by fins, or canted nozzles, that are attached to the motor.

15-15. The warhead is the portion of the rocket that produces the desired effect. It can be filled with explosives, toxic chemicals, WP, submunitions, CS, or illumination flares. The motor propels the rocket to the target. The fuse is the component that initiates the desired effect at the desired time. Rockets use the same type of fusing as projectiles and mortars. The fuse may be located in the nose or internally between the warhead and the motor.

15-16. Generally, the rocket motor will not create an additional hazard, because the motor is usually burned out shortly after the rocket leaves the launcher.

Guided Missiles

15-17. Guided missiles are like rockets, as they consist of the same parts; however, missiles are guided to their target by various guidance systems. Some of the smaller missiles, such as the tube-launched, optically tracked, wire-guided (TOW) and Dragon missiles, are wire-guided by the gunner to their targets.

15-18. Larger missiles, such as the phased-array, tracking radar intercept on target (PATRIOT) and the Sparrow are guided by radar to their target. Guided missiles use internal, proximity fusing and therefore, do not approach any guided missile you find laying on the battlefield.

Rifle Grenades

15-19. Rifle grenades look like mortars and are fired from a rifle that is equipped with a grenade launcher or an adapter. Many countries use rifle grenades as an Infantry direct-fire weapon. Some rifle grenades are propelled by specially designed blank cartridges, while others are propelled by standard ball cartridges. Rifle grenades may be filled with HEs, WP, CS, illumination flares, or colored screening smoke. They range in size from the small antipersonnel rifle grenade to the larger antitank rifle grenade. Antipersonnel rifle grenades use impact fusing. Some rifle grenades, such as the antitank version, have internal fusing behind the warhead; this type of fusing still functions on impact with the target.

THROWN ORDNANCE (HAND GRENADES)

15-20. Hand grenades are small items that may be held in one hand and thrown. All grenades have three main parts: a body, a fuse with a pull ring and safety clip assembly, and a filler. Never pick up a grenade you find on the battlefield, even if the spoon and safety pin are still attached. All grenades found laying on a battlefield should be considered booby-trapped. Thrown ordnance, commonly known as hand grenades, can be classified by use as follows:

- Fragmentation (also called defensive)
- Offensive
- Antitank
- Smoke
- Illumination

Fragmentation Grenades

15-21. Fragmentation grenades are the most common type of grenade and may be used as offensive or defensive weapons (see Figure 15-5). They have metal or plastic bodies that hold explosive fillers. These grenades produce casualties by high-velocity projection of fragments when they blow up. The fragmentation comes from the metal body or a metal fragmentation sleeve that can be internal or attached to the outside of the grenade. These grenades use a burning delay fuse that functions 3 to 5 seconds after the safety lever is released.

Offensive Grenades

15-22. Offensive grenades have a plastic or cardboard body and are not designed to have a lot of fragmentation. Their damage is caused from the over pressure of the explosive blast. These grenades use a burning-delay fuse that functions 3 to 5 seconds after the safety lever is released.

Antitank Grenades

15-23. Antitank grenades are designed to be thrown at tanks and other armored vehicles. They have a shaped-charge explosive warhead and are stabilized in flight by a spring-deployed parachute or a cloth streamer (Figure 15-6). These grenades use impact fuses.

SMOKE GRENADES

15-24. The two types of smoke grenades are bursting and burning (Figure 15-7). They may be made of rubber, metal, or plastic. Bursting-type smoke grenades are filled with WP and blow up when the fuse functions. These grenades use a burning delay fuse that functions 3 to 5 seconds after the safety lever is released. Burning-type smoke grenades produce colored smoke and use an instant-action fuse. There is no delay once the spoon is released. This is the same type of grenade that is used to dispense riot-control agents such as CS.

ILLUMINATION GRENADES

15-25. Illumination grenades are used for illuminating, signaling, and as an incendiary agent (Figure 15-8). The metal body breaks apart after the fuse functions and dispenses an illumination flare. This type of grenade uses a burning-delay fuse that functions 3 to 5 seconds after the safety lever is released.

Figure 15-5. Fragmentation grenades.

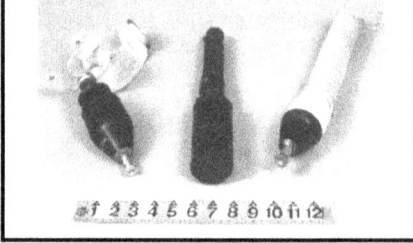

Figure 15-6. Antitank grenades.

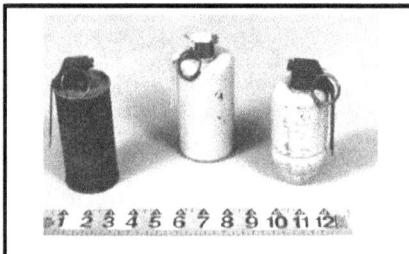

Figure 15-7. Smoke grenades.

Figure 15-8. U.S. illumination grenade.

Section II. IMPROVISED EXPLOSIVE DEVICES

IEDs are nonstandard explosive devices that target both Soldiers and civilians. IEDs range from crude homemade explosives to extremely intricate remote-controlled devices. They instill fear and diminish the resolve of our forces by escalating casualties. The sophistication and range of IEDs continue to increase as technology improves, and as our enemies gain experience.

TYPES

15-26. IEDs include explosive devices, impact-detonated devices, and vehicle-borne bombs:

TIMED EXPLOSIVE DEVICES

15-27. These can either be detonated by electronic means, possibly even by a cell phone; or by a combination of wire and either a power source or timed fuse.

IMPACT-DETONATED DEVICES

15-28. These detonate after any kind of impact such as after being dropped or thrown.

VEHICLE-BORNE BOMBS

15-29. Also known as car bombs, these explosive-laden vehicles are detonated via electronic command wire, wireless remote control, or a timed device(s). A driver is optional. Anything from a small sedan to a large cargo truck or cement truck (Figure 15-9) can be used. The size of the vehicle limits the size of the device. Bigger vehicles can carry much more explosive material, so they can cause more damage than smaller ones. Device functions also vary. Some possible signs of a car bomb include--

- A vehicle riding low, especially in the rear, and especially if the vehicle seems empty. However, because explosive charges can be concealed in the side panels, the weight may be distributed evenly. Even so, the vehicle may still ride low, indicating excessive weight.
- Large boxes, satchels, bags, or any other type of container in plain view such as on, under, or near the front seat of the vehicle.
- Wires or rope-like material coming out the front of the vehicle and leading to the rear passenger or trunk area.
- A timer or switch in the front of a vehicle. The main charge is usually out of sight, and as previously stated, often in the rear of the vehicle.
- Unusual or very strong fuel-like odors.
- An absent or suspiciously behaving driver.

Chapter 15

Vehicle Description	Maximum Explosives Capacity	Lethal Air Blast Range	Minimum Evacuation Distance	Falling Glass Hazard
Compact Sedan	500 Pounds 227 Kilos (In Trunk)	100 Feet 30 Meters	1,500 Feet 457 Meters	1,250 Feet 381 Meters
Full Size Sedan	1,000 Pounds 455 Kilos (In Trunk)	125 Feet 38 Meters	1,750 Feet 534 Meters	1,750 Feet 534 Meters
Passenger Van or Cargo Van	4,000 Pounds 1,818 Kilos	200 Feet 61 Meters	2,750 Feet 838 Meters	2,750 Feet 838 Meters
Small Box Van (14 Ft. box)	10,000 Pounds 4,545 Kilos	300 Feet 91 Meters	3,750 Feet 1,143 Meters	3,750 Feet 1,143 Meters
Box Van or Water/Fuel Truck	30,000 Pounds 13,636 Kilos	450 Feet 137 Meters	6,500 Feet 1,982 Meters	6,500 Feet 1,982 Meters
Semi-Trailer	60,000 Pounds 27,273 Kilos	600 Feet 183 Meters	7,000 Feet 2,134 Meters	7,000 Feet 2,134 Meters

Figure 15-9. Vehicle IED capacities and danger zones.

IDENTIFICATION

15-30. The following are tell-tale signs of IEDs:

- Wires
- Antennas
- Detcord (usually red in color)
- Parts of ordinance exposed

COMPONENTS

15-31. The following are components of an IED:

- Main Charge (Explosives) (Figure 15-10).
- Casing (material around the explosives; Figure 15-11).
- Initiators (command detonated, victim activated, and timer; Figure 15-12).

Unexploded Ordnance and Improvised Explosive Devices

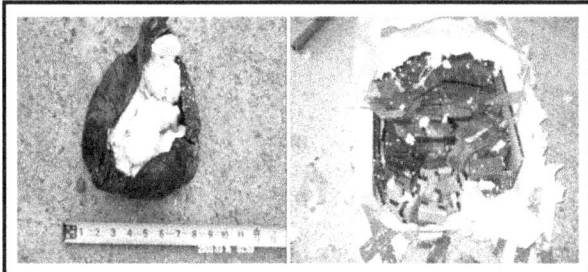

Figure 15-10. Main charge (explosives).

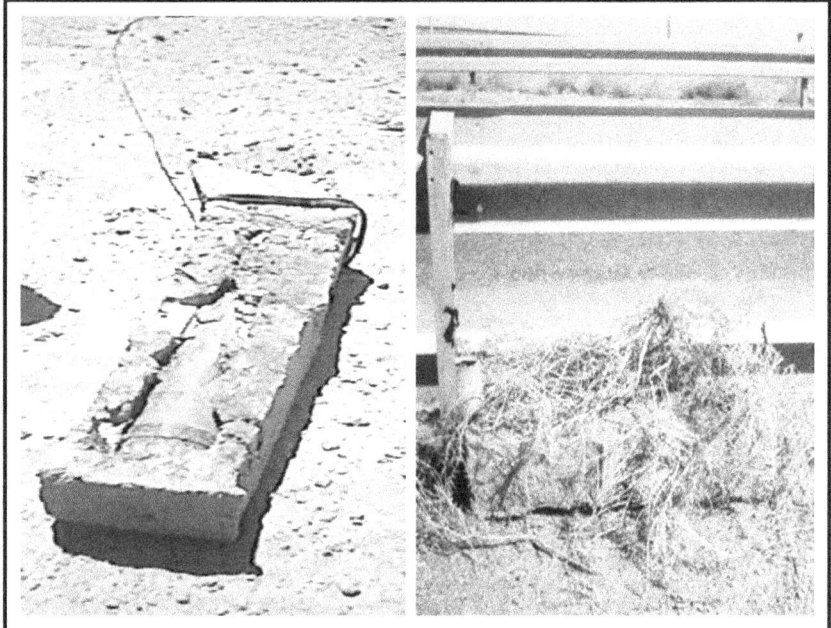

Figure 15-11. Casing (material around the explosives).

Chapter 15

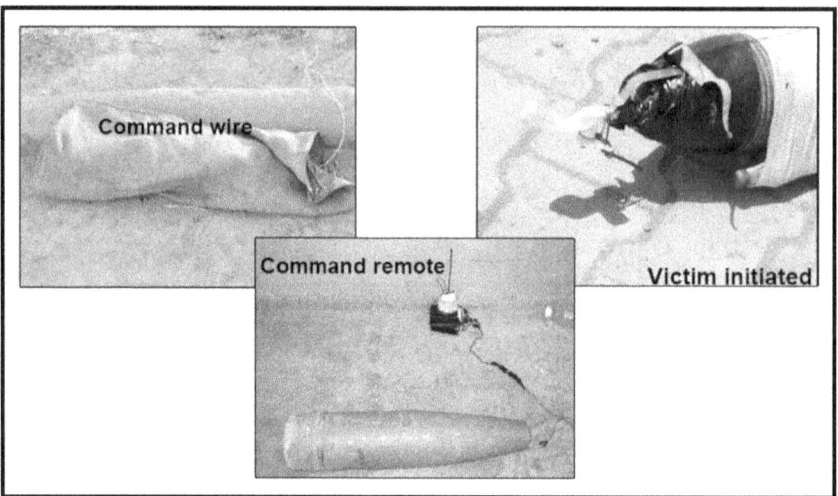

Figure 15-12. Initiators (command detonated, victim activated, with timer).

EXAMPLES

15-32. Figures 15-13 through 15-18, this page through page 15-16, show example IED types and components. These photos are examples to train Soldiers to recognize components of IEDs. Recognition is needed when Soldiers conduct operations, such as raids, traffic control points, convoys, and come across suspicious items.

Unexploded Ordnance and Improvised Explosive Devices

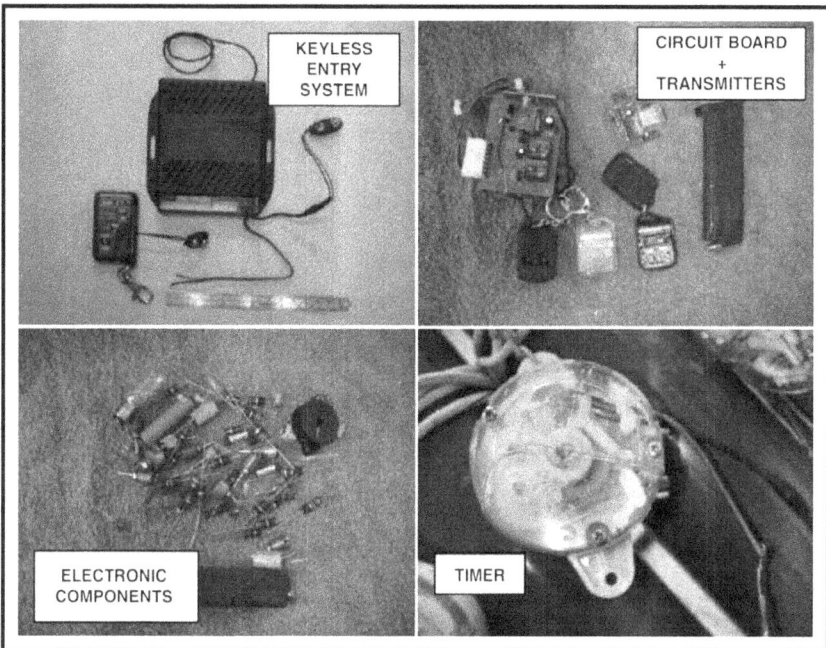

Figure 15-13. IED components.

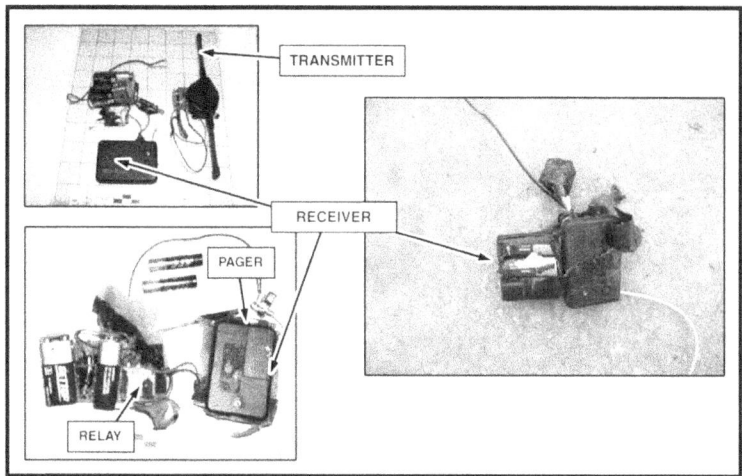

Figure 15-14. IED transmitters and receivers.

Chapter 15

Figure 15-15. Common objects as initiators.

Figure 15-16. Unexploded rounds as initiators.

Unexploded Ordnance and Improvised Explosive Devices

Figure 15-17. Emplaced IED with initiator.

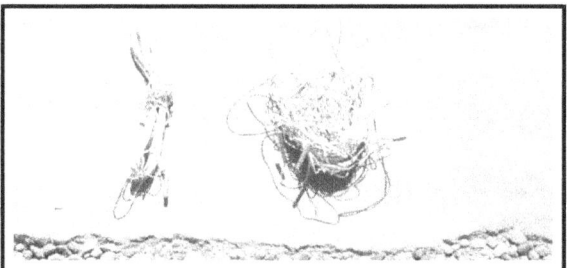

Figure 15-18. Electric blasting caps.

ACTIONS ON FINDING UXO

15-33. Many areas, especially previous battlefields, may be littered with a wide variety of sensitive and deadly UXO. Soldiers should adhere to the following precautions upon discovering a suspected UXO:

- Do not move toward the UXO. Some types of ordnance have magnetic or motion-sensitive fusing.
- Never approach or pick up UXO even if identification is impossible from a distance. Observe the UXO with binoculars if available.
- Send a UXO report (Figure 15-19) to higher HQ (see special segment below). Use radios at least 100 meters away from the ordnance. Some UXO fuses might be set off by radio transmissions.
- Mark the area with mine tape or other obvious material at a distance from the UXO to warn others of the danger. Proper markings will also help explosive ordnance disposal (EOD) personnel find the hazard in response to the UXO report.

Chapter 15

- Evacuate the area while carefully scanning for other hazards.
- Take protective measures to reduce the hazard to personnel and equipment. Notify local officials and people in the area.

Nine-Line UXO Incident Report

1. DTG: Date and time UXO was discovered.
2. Reporting Unit or Activity, and UXO Location: Grid coordinates.
3. Contact Method: How EOD team can contact the reporting unit.
4. Discovering Unit POC: MSE, or DSN phone number, and unit frequency or call sign.
5. Type of UXO: Dropped, projected, thrown, or placed, and number of items discovered.
6. Hazards Caused by UXO: Report the nature of perceived threats such as a possible chemical threat or a limitation of travel over key routes.
7. Resources Threatened: Report any equipment, facilities, or other assets threatened by the UXO.
8. Impact on Mission: Your current situation and how the UXO affects your status.
9. Protective Measures: Describe what you have done to protect personnel and equipment such as marking the area and informing local civilians.

Figure 15-19. Nine-Line UXO Incident Report.

ACTIONS ON FINDING IEDS

15-34. Follow these basic procedures when IEDs are found:

- Maintain 360-degree security. Scan close in, far out, high, and low.
- Move away. Plan for 300 meters distance minimum (when possible) and adapt to your METT-TC. Make maximum use of available cover. Get out of line of sight of IEDs.
- Always scan your immediate surroundings for more IEDs. Report additional IEDs to the on-scene commander.
- Try to confirm suspect IED. Always use optics. Never risk more than one person. Stay as far back as possible. When in doubt, back away and avoid touching.
- Cordon off the area. Direct people out of the danger area and do not allow anyone to enter besides those responsible for responding, such as EOD. Question, search, and detain suspects as needed. Check any and all locations that you move to for other IEDs.
- Report the situation to your higher command. Use the IED spot report shown in Figure 15-20.

IED SPOT REPORT

LINE 1. DATE-TIME-GROUP: [State when the item was discovered.]
LINE 2. UNIT:
LINE 3. LOCATION OF IED: [Describe as specifically as possible.]
LINE 4. CONTACT METHOD: [Radio frequency, call sign, POC.]
LINE 5. IED STATUS: [Detonation or no detonation.]
LINE 6. IED TYPE: [Disguised static / Disguised moveable / Thrown / Placed on TGT.]
LINE 7. NUMBER OF IEDs:
LINE 8. PERSONNEL STATUS:
LINE 6. EQUIPMENT STATUS:
LINE 7. COLLATERAL DAMAGE OR POTENTIAL FOR COLLATERAL DAMAGE:
LINE 8. TACTICAL SITUATION: [Briefly describe current tactical situation.]
LINE 9. REQUEST FOR: [QRF / EOD / MEDEVAC].
LINE 10. LOCATION OF L/U WITH REQUESTED FORCE (S):

Figure 15-20. IED Spot Report.

This page intentionally left blank.

Appendix A
Checklists and Memory Aids

This appendix consolidates all of the checklists in this publication. Other definitions Warrior Ethos, Soldier's Creed, Army Values, and more.

- Antipersonnel land mines, US national policy on
- Army Values
- Camouflage face paint, application to skin
- Code of Conduct
- Communication methods, comparison of
- Decontamination levels and techniques
- First aid
- IED Spot Report
- IED, vehicle, capacities and danger zones
- Illness in the field, rules for avoiding
- Individual fighting positions, characteristics of
- MOPP levels
- Nine-Line UXO Incident Report
- Personal predeployment checklist, example
- Potential indicators
- Prowords
- SALUTE format line by line
- Shelter checklist
- Soldier's Creed
- Survival
- Warrior Ethos

ANTIPERSONNEL LAND MINES, US NATIONAL POLICY ON

On May 16, 1996, the President of the United States implemented a phased restriction and elimination of antipersonnel land mines. Implementation began with non-self-destructing mines, but will eventually include all types of antipersonnel mines. This policy applies to all Infantry units either engaged in or training for operations worldwide. The use of non-self-destructing antipersonnel land mines is restricted to specific areas:

- Within internationally recognized national borders.
- In established demilitarized zones such as to defend South Korea.

Mines approved for use must be emplaced in an area with clearly marked perimeters. They must be monitored by military personnel and protected by adequate means to ensure the exclusion of civilians.

US policy also forbids US forces from using standard or improvised explosive devices as booby traps.

Except for South Korea-based units, and for units deploying to South Korea for training exercises, this policy forbids training with and employing inert M14 and M16 mines. This applies to units' home stations as well as at Combat Training Centers, except in the context of countermine or mine removal training.

- *Training with live M14 mines is UNAUTHORIZED!*
- *Training with live M16 mines is authorized only for Soldiers on South Korean soil.*

Exceptions:

This policy does not apply to standard use of antivehicular mines. Nor does it apply to training and using the M18 Claymore mine *in the command-detonated mode*.

When authorized by the appropriate commander, units may still use self-destructing antipersonnel mines such as the ADAM.

Authorized units may continue to emplace mixed minefields containing self-destructing antipersonnel land mines and antivehicular land mines such as MOPMS or Volcano.

The terms *mine, antipersonnel obstacle, protective minefield*, and *minefield* do not refer to an obstacle that contains non-self-destructing antipersonnel land mines or booby traps.

Any references to antipersonnel mines and the employment of minefields should be considered in the context of this policy.

ARMY VALUES

Loyalty	Bear true faith and allegiance to the Constitution, the Army, your unit, and other Soldiers.
Duty	Fulfill your obligations.
Respect	Treat people with dignity as they should be treated.
Selfless Service	Put the welfare of the nation, the Army, and your subordinates before your own.
Honor	Live up to all the Army Values.
Integrity	Do what's right, legally and morally.
Personal Courage (Physical or Moral)	Face fear, danger, or adversity.

CAMOUFLAGE FACE PAINT, APPLICATION TO SKIN

	SKIN COLOR	SHINE AREAS	SHADOW AREAS
CAMOUFLAGE MATERIAL	LIGHT OR DARK	FOREHEAD, CHEEKBONES, EARS, NOSE AND CHIN	AROUND EYES, UNDER NOSE, AND UNDER CHIN
LOAM AND LIGHT GREEN STICK	ALL TROOPS USE IN AREAS WITH GREEN VEGETATION	USE LOAM	USE LIGHT GREEN
SAND AND LIGHT GREEN STICK	ALL TROOPS USE IN AREAS LACKING GREEN VEGETATION	USE LIGHT GREEN	USE SAND
LOAM AND WHITE	ALL TROOPS USE ONLY IN SNOW-COVERED TERRAIN	USE LOAM	USE WHITE
BURNT CORK, BARK CHARCOAL, OR LAMP BLACK	ALL TROOPS, IF CAMOUFLAGE STICKS NOT AVAILABLE	USE	DO NOT USE
LIGHT-COLOR MUD	ALL TROOPS, IF CAMOUFLAGE STICKS NOT AVAILABLE	DO NOT USE	USE

Appendix A

CODE OF CONDUCT

I. I am an American, fighting in the forces which guard my country and our way of life. I am prepared to give my life in their defense.

II. I will never surrender of my own free will. If in command, I will never surrender the members of my command while they still have the means to resist.

III. If I am captured, I will continue to resist by all means available. I will make every effort to escape and aid others to escape. I will accept neither parole nor special favors from the enemy.

IV. If I become a prisoner of war, I will keep faith with my fellow prisoners. I will give no information or take part in any action which might be harmful to my comrades. If I am senior, I will take command. If not, I will obey the lawful orders of those appointed over me, and will back them up in every way.

V. When questioned, should I become a prisoner of war, I am required to give only name, rank, service number, and date of birth. I will evade answering further questions to the utmost of my ability. I will make no oral or written statements disloyal to my country (and its allies) or harmful to their cause.

VI. I will never forget that I am an American, fighting for freedom, responsible for my actions, and dedicated to the principles that made my country free. I will trust in my God and in the United States of America.

COMMUNICATION METHODS, COMPARISON OF

Method	Advantages	Disadvantages
Messengers	• Messengers are the most secure means of communication. • Messengers can hand carry large maps with overlays. • Messengers can deliver supplies along with messages. • Messengers are flexible (can travel long/short distances by foot or vehicle).	• Messengers are slow, especially if traveling on foot for a long distance. • Messengers might be unavailable, depending on manpower requirements (size of element delivering message). • Messengers can be captured by enemy.
Wire	• Wire reduces radio net traffic. • Wire reduces electromagnetic signature. • Wire is secure and direct. • Wire can be interfaced with a radio.	• Wire has to be carried (lots of it). • Wire must be guarded. • Wire is time consuming.
Visual Signals	• Visual signals aid in identifying friendly forces. • Visual signals allow transmittal of prearranged messages. • Visual signals are fast. • Visual signals provide immediate feedback.	• Visual signals can be confusing. • Visual signals are visible from far away. • The enemy might see them, too.
Sound	• Sound can be used to attract attention. • Sound can be used to transmit prearranged messages. • Sound can be used to spread alarms. • Everyone can hear it at once. • Sound provides immediate feedback.	• The enemy hears it also. • Sound gives away your position.
Radio	• Radios are the most frequently used means of communication. • Radios are fast. • Radios are light. • Radios can be interfaced with telephone wire.	• Radio is the least secure means of communication. • Radios require batteries. • Radios must be guarded or monitored.

DECONTAMINATION LEVELS AND TECHNIQUES

Levels	Techniques[1]	Purpose	Best Start Time	Performed By
Immediate	Skin decontamination Personal wipe down Operator wipe down Spot decontamination	Saves lives Stops agent from penetrating Limits agent spread Limits agent spread	Before 1 minute Within 15 minutes Within 15 minutes Within 15 minutes	Individual Individual or buddy Individual or crew Individual or crew
Operational	MOPP gear exchange[2] Vehicle wash down	Provides temporary relief from MOPP4 Limits agent spread	Within 6 hours Within 1 hour (CARC) or within 6 hours (nonCARC)	Unit Battalion crew or decontamination platoon
Thorough	DED and DAD DTD	Provides probability of long-term MOPP reduction	When mission allows reconstitution	Decontamination platoon Contaminated unit
Clearance	Unrestricted use of resources	METT-TC depending on the type of equipment contaminated	When mission permits	Supporting strategic resources

1. The techniques become less effective the longer they are delayed.
2. Performance degradation and risk assessment must be considered when exceeding 6 hours.

FIRST AID

1. Check for BREATHING.
2. Check for BLEEDING.
3. Check for SHOCK.

IED SPOT REPORT

LINE 1. DATE-TIME-GROUP: [State when the item was discovered.]
LINE 2. UNIT:
LINE 3. LOCATION OF IED: [Describe as specifically as possible.]
LINE 4. CONTACT METHOD: [Radio frequency, call sign, POC.]
LINE 5. IED STATUS: [Detonation or no detonation.]
LINE 6. IED TYPE: [Disguised static / Disguised moveable / Thrown / Placed on TGT.]
LINE 7. NUMBER OF IEDs:
LINE 8. PERSONNEL STATUS:
LINE 6. EQUIPMENT STATUS:
LINE 7. COLLATERAL DAMAGE OR POTENTIAL FOR COLLATERAL DAMAGE:
LINE 8. TACTICAL SITUATION: [Briefly describe current tactical situation.]
LINE 9. REQUEST FOR: [QRF / EOD / MEDEVAC].
LINE 10. LOCATION OF L/U WITH REQUESTED FORCE (S):

IED, VEHICLE, CAPACITIES AND DANGER ZONES

	Vehicle Description	Maximum Explosives Capacity	Lethal Air Blast Range	Minimum Evacuation Distance	Falling Glass Hazard
ATF	Compact Sedan	500 Pounds 227 Kilos (In Trunk)	100 Feet 30 Meters	1,500 Feet 457 Meters	1,250 Feet 381 Meters
	Full Size Sedan	1,000 Pounds 455 Kilos (In Trunk)	125 Feet 38 Meters	1,750 Feet 534 Meters	1,750 Feet 534 Meters
	Passenger Van or Cargo Van	4,000 Pounds 1,818 Kilos	200 Feet 61 Meters	2,750 Feet 838 Meters	2,750 Feet 838 Meters
	Small Box Van (14 Ft. box)	10,000 Pounds 4,545 Kilos	300 Feet 91 Meters	3,750 Feet 1,143 Meters	3,750 Feet 1,143 Meters
	Box Van or Water/Fuel Truck	30,000 Pounds 13,636 Kilos	450 Feet 137 Meters	6,500 Feet 1,982 Meters	6,500 Feet 1,982 Meters
	Semi-Trailer	60,000 Pounds 27,273 Kilos	600 Feet 183 Meters	7,000 Feet 2,134 Meters	7,000 Feet 2,134 Meters

ILLNESS IN THE FIELD, RULES FOR AVOIDING

- Never consume foods and beverages from unauthorized sources.
- Never soil the ground with urine or feces. Use a latrine or "cat hole."
- Keep your fingers and contaminated objects out of your mouth.
- Wash your hands--
  -- After any contamination.
  -- Before eating or preparing food.
  -- Before cleaning your mouth and teeth.
- Wash all mess gear after each meal or use disposable plastic ware once.
- Clean your mouth and teeth at least once each day.
- Avoid insect bites by wearing proper clothing and using insect repellents.
- Avoid getting wet or chilled unnecessarily.
- Avoid sharing personal items with other Soldiers, for example--
  -- Canteens.
  -- Pipes.
  -- Toothbrushes.
  -- Washcloths.
  -- Towels.
  -- Shaving gear.
- Avoid leaving food scraps lying around.
- Sleep when possible.
- Exercise regularly.

Appendix A

INDIVIDUAL FIGHTING POSITIONS, CHARACTERISTICS OF

Type	Position	Estimated Construction Time (man-hours)	Equipment Requirements	Direct Small Caliber Fire	Indirect Fire Blast and Fragmentation (Near-Miss)*	Indirect Fire Blast and Fragmentation (Direct Hit)	Nuclear Weapons**	Remarks
Hasty	Crater	0.2	Hand tools	7.62 mm	Better than in open – no overhead protection	None	Fair	
Hasty	Skirmisher's trench	0.5	Hand tools	7.62 mm	Better than in open – no overhead protection	None	Fair	
Hasty	Prone position	1.0	Hand tools	7.62 mm	Better than in open – no overhead protection	None	Fair	Provides all-around cover
Deliberate	One-soldier position	3.0	Hand tools	12.7 mm	Medium artillery no closer than 30 ft – no overhead protection	None	Fair	
Deliberate	One-soldier position with 1.5 ft overhead cover	8.0	Hand tools	12.7 mm	Medium artillery no closer than 30 ft	None	Good	Additional cover provides protection from direct hit small mortar blast
Deliberate	Two-soldier position	6.0	Hand tools	12.7 mm	Medium artillery no closer than 30 ft – no overhead protection	None	Fair	
Deliberate	Two-soldier position with 1.5 ft overhead cover	11.0	Hand tools	12.7 mm	Medium artillery no closer than 30 ft	None	Good	Additional cover provides protection from direct hit small mortar blast
Deliberate	AT-4 position	3.0	Hand tools	12.7 mm	Medium artillery no closer than 30 ft – no overhead protection	None	Fair	

Note: Chemical protection is assumed because of individual protective masks and clothing.
* Shell sizes are: Small Medium
  Mortar 82mm 120mm
  Artillery 105mm 152mm
** Nuclear protection ratings are rated poor, fair, good, very good, and excellent.

MOPP LEVELS

Equipment	MOPP Ready	MOPP0	MOPP1	MOPP2	MOPP3	MOPP4	Mask Only
Mask	Carried	Carried	Carried	Carried	Worn	Worn	Worn***
JSLIST	Ready*	Avail**	Worn	Worn	Worn	Worn	
Overboots	Ready*	Avail**	Avail**	Worn	Worn	Worn	
Gloves	Ready*	Avail**	Avail**	Avail**	Avail**	Worn	
Helmet Cover	Ready*	Avail**	Avail**	Worn	Worn	Worn	

* Item must be available to Soldier within two hours, with replacement available within six hours.
** Item must be positioned within arm's reach of the Soldier.
*** Soldier Never "mask only" if a nerve or blister agent has been used in the AO.

NINE-LINE UXO INCIDENT REPORT

1. DTG: Date and time UXO was discovered.
2. Reporting Unit or Activity, and UXO Location: Grid coordinates.
3. Contact Method: How EOD team can contact the reporting unit.
4. Discovering Unit POC: MSE, or DSN phone number, and unit frequency or call sign.
5. Type of UXO: Dropped, projected, thrown, or placed, and number of items discovered.
6. Hazards Caused by UXO: Report the nature of perceived threats such as a possible chemical threat or a limitation of travel over key routes.
7. Resources Threatened: Report any equipment, facilities, or other assets threatened by the UXO.
8. Impact on Mission: Your current situation and how the UXO affects your status.
9. Protective Measures: Describe what you have done to protect personnel and equipment such as marking the area and informing local civilians.

PERSONAL PREDEPLOYMENT CHECKLIST, EXAMPLE

Defense Enrollment Eligibilty Reporting System (DEERS)	Verify your DEERS information and ensure your family members can get needed medical care in your absence.
Dental Records	Update if needed.
Documents, Locations of	Ensure that your spouse or other family member(s) know where to find all of the above documents.
DD Form 93 Emergency Data Record	Check to ensure this is current and correct.
Eyeglasses and Protective Mask Inserts	Ensure you have two pairs of eyeglasses and protective mask inserts, all with your current prescription, if required.
Family Assistance	
Army Community Service (ACS)	Tell your family where to get various kinds of support and help while you are gone.
Legal Aid, Military	Tell your spouse where to get military legal aid in your absence.
Readiness Group (FRG)	Tell your family where to get various kinds of support and help while you are gone.
Finances	Ensure your spouse has access to all of your records and accounts and update them as needed.
Identification	Ensure you have two sets of these, if required.
Legal	See Family Assistance.
Life Insurance	Designate your beneficiary on SGLV Forms 8286 and 8286A, Soldier's Group Life Insurance (SGLI) Election and Certificate.
Directive, Advance	Specify any decisions you wish others to make on your behalf should you be unable to do so for yourself.
Medical (see also Power of Attorney)	
DD Form 2766 (Shot Record)	Keep your vaccinations and immunizations current.
Directive, Advance	Prepare if you want to specify how decisions are made on your behalf should you be unable to do so for yourself.
Living Will	Prepare if desired.
Records	Update if needed.
Warnings Tags	Ensure you have two sets of these, if required.
Power of Attorney	
General	Prepare to allow someone to perform all duties for you in your absence.
Medical, Durable	Prepare one of these to designate who makes decisions for you or your dependents, including your minor children, should a medical emergency occur while you are deployed or otherwise unable to make the decision yourself.
Special	Prepare to allow someone to perform a particular kind of duty for you in your absence.
Property	Prepare or update accounts, documents, and records as needed.
Service Record	Check to ensure this is current and correct.
Training	Update your weapon qualification(s), if needed.
Will(s)	Prepare new or update existing, for you and your spouse, if needed.

POTENTIAL INDICATORS

SIGHT Look for--	SOUND Listen for--	TOUCH Feel for--	SMELL Smell for--
• Enemy personnel, vehicles, and aircraft • Sudden or unusual movement • New local inhabitants • Smoke or dust • Unusual movement of farm or wild animals • Unusual activity--or lack of activity--by local inhabitants, especially at times or places that are normally inactive or active • Vehicle or personnel tracks • Movement of local inhabitants along uncleared routes, areas, or paths • Signs that the enemy has occupied the area • Evidence of changing trends in threats • Recently cut foliage • Muzzle flashes, lights, fires, or reflections • Unusual amount (too much or too little) of trash	• Running engines or track sounds • Voices • Metallic sounds • Gunfire, by weapon type • Unusual calm or silence • Dismounted movement • Aircraft	• Warm coals and other materials in a fire • Fresh tracks • Age of food or trash	• Vehicle exhaust • Burning petroleum products • Food cooking • Aged food in trash • Human waste

OTHER CONSIDERATIONS

Armed Elements	Locations of factional forces, mine fields, and potential threats.
Homes and Buildings	Condition of roofs, doors, windows, lights, power lines, water, sanitation, roads, bridges, crops, and livestock.
Infrastructure	Functioning stores, service stations, and so on.
People	Numbers, gender, age, residence or DPRE status, apparent health, clothing, daily activities, and leadership.
Contrast	Has anything changed? For example, are there new locks on buildings? Are windows boarded up or previously boarded up windows now open, indicating a change in how a building is expected to be used? Have buildings been defaced with graffiti?

PROWORDS

PROWORD	MEANING
ALL AFTER	I refer to the entire message that follows...
ALL BEFORE	I refer to the entire message that precedes...
BREAK	I now separate the text from other parts of the message.
CORRECTION	There is an error in this transmission. This will continue with the last word correctly transmitted.
GROUPS	This message contains the number of groups indicated by the numeral following.
I SAY AGAIN	I am repeating transmission or part indicated.
I SPELL	I shall spell the next word phonetically.
MESSAGE	A message that requires recording is about to follow. (Transmitted immediately after the call.) This proword is not used on nets primarily employed for conveying messages. It is intended for use when messages are passed on tactical or reporting net.
MORE TO FOLLOW	Transmitting station has additional traffic for the receiving station.
OUT	This is the end of my transmission to you and no answer is required or expected.
OVER	This is the end of my transmission to you and a response is necessary. Go ahead: transmit.
RADIO CHECK	What is my signal strength and readability, i.e. How do you hear me?
ROGER	I have received your last transmission satisfactorily, radio check is loud and clear.
SAY AGAIN	Repeat all of your last transmission. Followed by identification data means "repeat - (portion indicated)."
THIS IS	This transmission is from the station whose designator immediately follows.
TIME	That which immediately follows is the time or date-time group of the message.
WAIT	I must pause for a few seconds.
WAIT-OUT	I must pause longer than a few seconds.
WILCO	I have received your transmission, understand it, and will comply, to be used only by the addressee. Since the meaning of ROGER is included in that of WILCO, the two prowords are never used together.
WORD AFTER	I refer to the word of the message that follows.
WORD BEFORE	I refer to the word of the message that precedes.

SALUTE

Line No.	Type Info	Description
1	Size / Who	Expressed as a quantity and echelon or size. For example, report "10 enemy Infantrymen" (not "a rifle squad").
	If multiple units are involved in the activity you are reporting, you can make multiple entries.	
2	Activity / What	Relate this line to the PIR being reported. Make it a concise bullet statement. Report what you saw the enemy doing, for example, "emplacing mines in the road."
3	Location / Where	This is generally a grid coordinate, and should include the 100,000-meter grid zone designator. The entry can also be an address, if appropriate, but still should include an eight-digit grid coordinate. If the reported activity involves movement, for example, advance or withdrawal, then the entry for location will include "from" and "to" entries. The route used goes under "Equipment/How."
4	Unit / Who	Identify who is performing the activity described in the "Activity/What" entry. Include the complete designation of a military unit, and give the name and other identifying information or features of civilians or insurgent groups.
5	Time / When	For future events, give the DTG for when the activity will initiate. Report ongoing events as such. Report the time you saw the enemy activity, not the time you report it. Always report local or Zulu (Z) time.
6	Equipment / How	Clarify, complete, and expand on previous entries. Include information about equipment involved, tactics used, and any other essential elements of information (EEI) not already reported in the previous lines.

SHELTER CHECKLIST

- B Blend
- L Low silhouette
- I Irregular shape
- S Small
- S Secluded location

Appendix A

SOLDIER'S CREED

I am an American Soldier.
I am a Warrior and a member of a team.
I serve the people of the United States and live the Army Values.
I will always place the mission first.
I will never accept defeat.
I will never quit.
I will never leave a fallen comrade.
I am disciplined, physically and mentally tough, trained, and proficient in my Warrior tasks and drills.
I always maintain my arms, my equipment, and myself.
I am an expert, and I am a professional.
I stand ready to deploy, engage, and destroy the enemies of the United States of America in close combat.
I am a guardian of freedom and the American way of life.
I am an American Soldier.

SURVIVAL

S	**S**ize up the situation (surroundings, physical condition, equipment)
U	**U**se all your senses. Undue haste makes waste
R	**R**emember where you are
V	**V**anquish fear and panic
I	**I**mprovise
V	**V**alue living
A	**A**ct like the natives
L	**L**ive by your wits, but for now **L**earn basic skills

WARRIOR ETHOS

I will always place the mission first.
I will never accept defeat.
I will never quit.
I will never leave a fallen comrade.

Glossary

Section I. ACRONYMS AND ABBREVIATIONS

1SG	first sergeant

A

AAR	after-action review
AC	alternating current
ACADA	automatic chemical agent decontaminating apparatus
ACH	advanced combat helmet
ACS	Army community service
ACU	Army combat uniforms
AMAT	anti-materiel
ANCD	automated net control device
AO	area of operation
AP	antipersonnel
APOBS	Antipersonnel Obstacle Breaching System
ARNG	Army National Guard
ARNGUS	Army National Guard of the United States
ASIP	advanced system improvement program
AT	antitank
ATNAA	antidote treatment, nerve agent, auto injector

B

BDM	bunker defeat munition
BDO	battle dress overgarment
BIS	backup iron sight
BLISS	blend, low, silhouette, irregular, shape, small, secluded location
BVO	black vinyl overshoe

C

C	Celsius (degrees)
CANA	convulsant antidote for nerve agents
C-A-T	combat application tourniquet
CB	chemical and biological
CBRN	chemical, biological, radiological, or nuclear
CCIR	commander's critical information requirements
CCM	close combat missiles
CCO	close combat optic
CED	captured enemy document
CEE	captured enemy equipment
CLP	cleaner lubricant preservative
CLU	command launch unit
COMSEC	communications security
CPFC	chemical protective footwear cover
CPHC	chemical protective helmet cover
CS	confined space
CW	chemical warfare
CVC	combat vehicle crew

D

DAP	decontamination apparatus portable
DED	detailed equipment decontamination
DEERS	Defense Enrollment Eligibility Reporting System
DEET	N-diethyl-m-toluamide
det	detonator
DOD	Department of Defense
DPRE	displaced persons, refugees, or evacuees
DTD	detailed troop decontamination

E

EEFI	essential elements of friendly information
EEI	essential elements of information
EMP	electromagnetic pulse
EOD	explosive ordnance disposal
EPA	evasion plan of action
EPW	enemy prisoner of war
ES2	'Every Soldier is a Sensor' concept

F

F	Fahrenheit (degrees)
FASCAM	family of scatterable mines
FOUO	for official use only
FOV	field of view
FP	force protection
FPL	final protective line

Glossary

FSG	family support group

G

G-2	assistant chief of staff for intelligence
GPFU	gas particulate filter unit
GVO	green vinyl overshoe

H

HE	high explosive
HEAT	high explosive antitank
HEDP	high-explosive dual purpose
HP	high penetration
HUMINT	human intelligence
HWTS	heavy weapon thermal sight

I

IAW	in accordance with
IBA	interceptor body armor
ICAM	improved chemical agent monitor
ID	identification
IED	improvised explosive device
IFAK	improved first-aid kit
IHFR	improved high frequency radio
IMT	individual movement technique
IR	infrared

J

JSLIST	joint service lightweight integrated suit technology

K

kph	kilometer per hour

L

LAW	light antiarmor weapon
lb	pound
LWTS	light weapon thermal sight

M

m	meter
max	maximum
MBITR	multiband intrateam radio
MDI	modernized demolition initiator
MEL	maximum engagement line
METT-TC	mission, enemy, terrain, troops, and equipment, time available, and civil considerations

MHz	megahertz
mm	millimeter
MOLLE	modular lightweight load-carrying equipment
MOPMS	Modular Pack Mine System
MOPP	mission-oriented protective posture
MOS	military occupational specialty
mph	miles per hour
MTF	medical treatment facility
MULO	multipurpose vinyl overshoe
MWTS	medium weapon thermal sight

N

NAAK	nerve agent antidote kit
NATO	North Atlantic Treaty Organization
NBC	nuclear, biological, chemical (obsolete; see CBRN)
NCO	noncommissioned officer
NSN	national stock number
NVD	night vision device
NVG	night vision goggles

O

OD	olive drab
OHC	overhead cover
OOTW	operations other than war
OP	observation post
OPORD	operation order
OPSEC	operations security
OTV	outer tactical vest

P

PATRIOT	phased array, tracking radar intercept on target
PDF	principal direction of fire
PIR	priority intelligence requirement
POW	prisoner of war

R

RCU	radio control unit
RDD	radiological dispersal device
RF	radio frequency
ROE	Rules of Engagement
RP	reference point
RS	reduced sensitivity
RTU	receiver transmitter unit

Glossary

S

S-2	battalion/brigade intelligence officer
SABA	self-aid/buddy-aid
SALUTE	size, activity, location, uniform, time, and equipment
SATCOM	single-channel tactical satellite communications
SAW	squad automatic weapon
SCPE	simplified collective protection equipment
SDS	Sorbent Decontamination System
SERE	survival, evasion, resistance, and escape
SGLI	Soldier's Group Life Insurance
SINCGARS	Single-Channel Ground and Airborne Radio System
SIP	system improvement program
SLLS	stop, look, listen, smell
SLM	shoulder-launched munition
SMAW-D	shoulder-launched, multipurpose, assault- weapon disposable
SOI	signal operating instructions
SOP	standing operating procedures
SRTA	short-range training ammunition

T

TIB	toxic industrial biological
TIC	toxic industrial chemical
TIM	toxic industrial material
TIR	toxic industrial radiological
TM	technical manual
TOW	tube-launched, optically tracked, wire-guided
TP	training practice
TRP	target reference point
TWS	thermal weapon sight

U

US	United States
USAIS	United States Army Infantry School
USAR	United States Army Reserve
UXO	unexploded ordnance

W

WBD	warrior battle drill
WFOV	wide field of view
WP	white phosphorous
WRP	weapon reference point

Section II. TERMS

A

arroyo	steep-walled, eroded valley; same as 'wadi'

C

camouflage	protection from identification
concealment	protection from observation only
cover	protection from weapons fire, explosions, fragments, flames, CBRN effects, and observation

F

flag	to allow a weapon to extend beyond the corner of a building

G

gebel	mountain or mountain range

I

indicator	information, needed by the commander to make decisions, on the intention or capability of a potential enemy

M

mirage	an optical phenomenon caused by the refraction of light through heated air rising from a sandy or stony surface

N

nipa palm	a creeping, semiaquatic palm whose sap is a source of nipa fruit and of sugar, whose seeds are edible, and whose long, strong leaves are used in thatching and basketry

O

overhead cover protects Soldier from indirect fire

P

parapet enables Soldier to engage enemy within assigned sector of fire while protecting the soldier from direct fire

pie-ing aiming a weapon beyond the corner of a building in the direction of travel, without allowing the weapon to extend beyond the corner, and then side-stepping around the corner in a circular fashion with the muzzle of the weapon as the pivot point

S

sago palm a tall palm with long leaves that curved backward, inward, or downward, and whose porous trunk is ground and used to thicken foods and stiffen textiles

savanna a temperate grassland with scattered trees

skirmisher's trench a shallow ditch used as a hasty fighting position

spoil excavated earth

T

toxic industrial materials (TIMs) includes toxic industrial chemical, biologica and radioactive materials; are produced to prescribed toxicity levels; are administered through inhalation (mostly), ingestion. or absorption; may be stored or used in any environment fo any tactical purpose--medical. industrial, commercial, military, or domestic. MOPP gear may or may not protect against TIMs.

W

wadi steep-walled, eroded valley; same as 'arroyo'

warrior ethos four items extracted from the middle of the Soldier's Creed:
1. I will always place the mission first.
2. I will never accept defeat
3. I will never quit
4. I will never leave a fallen comrade.

wind chill the effect of moving air on exposed flesh

References

SOURCES USED/DOCUMENTS NEEDED

These are the sources quoted or paraphrased in this publication. All must be available to the intended users of this publication.

ARMY REGULATION
AR 525-28, *Personnel Recovery*, 5 March 2010.

ARMY TECHNIQUES PUBLICATION
ATP 3-11.37, *Multi-Service Tactics, Techniques, and Procedures for Nuclear, Biological, and Chemical Reconnaissance and Surveillance*, 25 March 2013.

ATP 3-37.34, *Survivability Operations*, 28 June 2013.

ARMY TACTICS, TECHNIQUES, AND PROCEDURES
ATTP 3-06.11, *Combined Arms Operations in Urban Terrain*, 10 June 2011.

ATTP 3-21.9, *SBCT Infantry Rifle Platoon and Squad*, 8 December 2010.

FIELD MANUALS
FM 2-0, *Intelligence*, 23 March 2010.

FM 3-05.70, *Survival*, 17 May 2002.

FM 3-11.3, *Multiservice Tactics, Techniques, and Procedures for Chemical, Biological, Radiological, and Nuclear Contamination Avoidance*, 2 February 2006.

FM 3-11.4, *Multiservice Tactics, Techniques, and Procedures for Nuclear, Biological, and Chemical (NBC) Protection*, 2 June 2003.

FM 3-11.5, *Multiservice Tactics, Techniques, and Procedures for Chemical, Biological, Radiological, and Nuclear Decontamination*, 4 April 2006.

FM 3-21.8, *The Infantry Platoon and Squad*, 28 March 2007.

FM 3-22.9, *Rifle Marksmanship M16- M4-Series Weapons*, 12 August 2008.

FM 3-22.27, *MK 19, 40-mm Grenade Machine Gun, Mod 3*, 28 November 2003.

FM 3-22.37, *Javelin--Close Combat Missile System, Medium*, 20 March 2008.

FM 3-22.65, *Browning Machine Gun, Caliber .50 HB, M2*, 3 March 2005.

FM 3-22.68, *Crew Served Weapons*, 21 July 2006.

FM 3-23.35, *Combat Training with Pistols, M9, and M11*, 25 June 2003.

FM 3-34.210, *Explosive Hazards Operations*, 27 March 2007.

FM 3-34.214, *Explosives and Demolitions*, 11 July 2007.

FM 4-02.2, *Medical Evacuation*, 8 May 2007.

FM 4-02.7, *Tactics, Techniques and Procedures for Health Service Support in a Chemical, Biological, Radiological, and Nuclear*, 15 July 2009.

FM 4-02.285, *Multiservice Tactics, Techniques and Procedures for Treatment of Chemical Agent Casualties and Conventional Military Chemical Injuries*, 18 September 2007.

FM 4-25.11, *First Aid*, 23 December 2002.

FM 6-22, *Army Leadership: Competent, Confident, and Agile*, 12 October 2006.

FM 6-30, *Tactics, Techniques, and Procedures for Observed Fire*, 16 July 1991.

FM 7-21.13, *The Soldier's Guide*, 2 February 2004.

FM 8-10-6, *Medical Evacuation in a Theater of Operations Tactics, Techniques, and Procedures*, 14 April 2000.

References

> **Note:** FM 8-10-6, Chapters 1-7 and Appendixes A, B, D through F, K, L, and N were superseded by FM 4-02.2, 8 May 2007.

FM 27-10, *The Law of Land Warfare*, 15 July 1976.

FORMS
DA Form 2028, *Recommend Changes to Publications and Blank Forms.*
DA Form 5517-R, *Standard Range Card (LRA).*
DA Form 7425, *Readiness and Deployment Checklist.*
DD Form 93, *Record of Emergency Data.*
DD Form 2745, *Enemy Prisoner of War (EPW) Capture Tag.*
DD Form 2766, *Adult Preventive and Chronic Care Flowsheet.*
SGLV Form 8286, *Servicemembers' Group Life Insurance Election and Certificate.*
SGLV Form 8286A, *Family Coverage Election.*

STUDENT HANDBOOK
SH 3-22.9, *Sniper Data Book.*

SOLDIER TRAINING PUBLICATIONS
STP 7-11B1-SM-TG, *Soldier's Manual and Trainer's Guide MOS 11B Infantry, Skill Level 1*, 6 August 2004.
STP 21-1-SMCT, *Soldier's Manual of Common Tasks, Skill Level 1*, 11 September 2012.

TECHNICAL MANUALS
TM 3-22.31, *40-mm Grenade Launchers*, 17 November 2010.
TM 3-23.25, *Shoulder-Launched Munitions*, 15 September 2010.
TM 9-1345-209-10, *Operator's Manual for Modular Pack Mine System (MOPMS) Consisting of Dispenser and Mine, Ground: M131 (NSN 1345-01-160-8909) Control, Remote, Land Mine System: M71 (1290-01-161-3662) and Dispenser and Mine, Ground, Training: M136 (6920-01-162-9380)*, 31 March 1992.
TM 9-1375-213-12, *Operator's and Unit Maintenance Manual (Including Repair Parts and Special Tools List): Demolition Materials*, 30 March 1973.
TM 9-1375-219-13&P, *Operator's, Unit and Direct Support Maintenance Manual for Demolition Kit, Breaching System, Anti-Personnel Obstacle (APOBS) (DODIC, MN79) (NSN 1375-01-426-1376) and Dummy Demolition Kit, Breaching System, Anti-Personnel Obstacle (APOBS) (DODIC MN84) (1375-01-467-1277)*, 4 June 2004.
TM 11-5855-301-12&P, *Operator's and Unit Maintenance Manual (Including Repair Parts and Special Tools List) for Light, Aiming, Infrared AN/PAQ-4B (NSN 5855-01-361-1362) (EIC: N/A) AN/PAQ-4C (5855-01-398-4315) (EIC: N/A)*, 15 May 2000.

INTERNET WEB SITES
Some of the documents listed in these references may be downloaded from Army websites. A few require AKO login:

Air Force Pubs	http://afpubs.hq.af.mil/
Army Human Resources	https://www.hrc.army.mil
Army Forms	https://armypubs.us.army.mil/eforms/index.html
Army Knowledge Online	https://armypubs.us.army.mil/doctrine/index.html
Army Publishing Directorate	http://www.apd.army.mil/
NATO ISAs	http://www.nato.int/docu/standard.htm
Reimer Digital Library	http://www.train.army.mil

Index

10-meter boresight target, 10-25 (*illus*)

A

abdominal thrust, 3-10 (*illus*)
advanced camouflage face paint, 5-9 (*illus*)
Advanced System Improvement Program (ASIP), 11-14 (*illus*)
aerial flares, 7-8 (*illus*)
aim point, 10-36
airway, 3-4 (*illus*)
alternate fireman's carry, 3-32 (*illus*)
AN/PAQ-4, 9-14 (*illus*)
AN/PAS-13, V1 through V3, 9-13 (*illus*)
AN/PEQ-2A, 9-14 (*illus*)
AN/PRC-119A-D, 11-13 (*illus*)
AN/PRC-119E Advanced System Improvement Program (ASIP), 11-14 (*illus*)
AN/PRC-148 multiband intrateam radio (MBITR), 11-12 (*illus*)
AN/PVS-14, 9-12 (*illus*)
AN/PVS-7, 9-12 (*illus*)
anti-materiel/antitank submunitions (conventional), 15-5 (*illus*)
antipersonnel/anti-materiel submunitions (conventional), 15-4 (*illus*)
antipersonnel, ball type submunitions, 15-3 (*illus*)
antipersonnel mines, 14-2
M18A1, 14-3 (*illus*)
antitank grenades, 15-9 (*illus*)
AN/UDR-13 radiac set, 13-21 (*illus*)
AN/VDR-2 radiac set, 13-21 (*illus*)
appearance-of-objects method, 9-16 (*illus*)
arctic environment, 4-8
area denial submunitions (conventional), 15-3 (*illus*)
Army Values, 1-3 (*illus*)

ASIP. *See* Advanced System Improvement Program
assessment of casualty, 3-3 (*illus*)
automated net control device (ANCD), 11-9 (*illus*)
automatic chemical agent decontaminating apparatus (ACADA), M22, 13-5 (*illus*)

B

backup iron sight, 10-25 (*illus*)
bandages, 3-15 through 3-16 (*illus*)
bandoleer, M7, 14-3 (*illus*)
band pulled tight, 3-22 (*illus*)
bangalore torpedo, 14-18 through 14-19 (*illus*)
basement windows, passing, 8-3 (*illus*)
battlesight zero, 10-18
biological weapons, 13-16
blasting caps, electric, 15-16 (*illus*)
body armor, 5-3 (*illus*)
booster assemblies, 14-9
borelight, 10-23 (*illus*)
borelight zero, 10-21
boresight target and offset symbols, 10-25 (*illus*)
breach and cross
minefield, 14-14
wire obstacle, 14-18
breaching procedures, 14-1
breastbone depressed 1 1/2 to 2 inches, 3-12 (*illus*)
building
entering, 8-6
moving within, 8-5
built-down overhead cover, 6-18 (*illus*)

C

C4
detonating cord, 14-11 (*illus*)
L-shaped charge, 14-11 (*illus*)
modernized demolition initiator, 14-10 (*illus*)
call signs, 11-10 (*illus*)

camouflage, 5-5, 6-4. *See also* fighting positions
continued improvement, 6-3 (*illus*)
face paint, 5-10 (*illus*)
helmet, 5-8 (*illus*)
Soldiers, 5-7 (*illus*)
captured document tag, 9-6 (*illus*)
casing (material around the explosives), 15-13 (*illus*)
casualty assessment, 3-3 (*illus*)
casualty evacuation (CASEVAC), 3-28
check for breathing, 3-6 (*illus*)
chemical agent
detector kit, M256, 13-8 (*illus*)
detector paper, 13-6 thru 13-7 (*illus*)
detectors, 13-6 through 13-8 (*illus*)
chemical, biological, radiological, or nuclear (CBRN) weapons, 13-1
chest thrust, 3-11 (*illus*)
classifications, 11-8
clearing of room, 8-11
close combat missile fighting positions, 6-24
clothing and sleeping gear, 3-40
Code of Conduct, 12-7 (*illus*)
colors, 5-6
combat lifesaver, 3-1
combat marksmanship, 10-1
combat zero, 10-16
common messages, 11-5
common objects as initiators, 15-15 (*illus*)
common prowords, 11-7 (*illus*)
communication methods, 11-2 (*illus*)
communications, 11-1
communications security, 11-8
concealment, 5-4 (*illus*), 6-3
See also fighting positions

Index

convulsant antidote for nerve agents (CANA), 13-2 (*illus*)
corners, moving around, 8-4 (*illus*)
cover, 5-1, 6-2 (*illus*)
  along a wall, 6-1. *See also* fighting positions
  natural, 5-1 (*illus*)
cradle drop drag, 3-35 (*illus*)
crossed-finger method, 3-13 (*illus*)

D

DA Form 7425, *Readiness and Deployment Checklist*, 2-1
data section, 6-29 through 6-30 (*illus*)
DD Form 2745, *Enemy Prisoner of War (EPW) Capture Tag*, 9-7
decontamination, 13-12
  levels and techniques, 13-13 (*illus*)
demolitions, 14-1, 14-9
desert, 4-1
dislodging of foreign body, 3-13 (*illus*)
dispersion, 5-7
dressing, 3-17 through 3-19 (*illus*)
drills, 1-4
  battle drills, 1-4
  warrior drills, 1-5
dropped ordnance, 15-1

E

electric blasting caps, 15-16 (*illus*)
elevation of injured limb, 3-18 (*illus*)
emergency bandage, 3-15 (*illus*)
emplacement of machine gun in a doorway, 8-18 (*illus*)
entering a building, 8-6
equipment decontamination kit, M295, 13-15 (*illus*)
escape, 12-8
establishment of sectors and building method, 6-9 (*illus*)
evasion, 12-4
Every Soldier is a Sensor, 9-1

F

face paint, 5-10 (*illus*)
feet, 3-40 (*illus*)
fighting positions, 6-1, 8-13
  one-man, 6-20
final protective line (FPL), 6-27 (*illus*)
final shot group results, 10-20 (*illus*)
fire and movement, 7-9
fire control, 10-14
fireman's carry, 3-30 (*illus*)
  alternate, 3-32 (*illus*)
fire team wedge, 7-5 (*illus*)
firing platforms, 6-21 (*illus*)
first aid, 3-2 (*illus*)
foreign body, dislodging, 3-13 (*illus*)
forms. *See* form type
fragmentation grenades, 15-9 (*illus*)

G

grenade sump, 6-14 (*illus*), 6-22 (*illus*)
ground flares, 7-7 (*illus*)

H

handling and reporting of the enemy, 9-6
hasty and deliberate fighting positions, 6-5
hasty fighting position, 8-14
head tilt/chin lift technique, 3-5 (*illus*)
helmet, 5-3 (*illus*)

I

IC-F43 portable UHF transceiver, 11-13 (*illus*)
IED Spot Report, 15-18 (*illus*)
IFAK. *See* improved first-aid kit
illness, rules for avoiding, 3-39 (*illus*)
illumination grenade, 15-9 (*illus*)
immediate action, 10-31
  while moving, 7-6
impact area, 7-6 (*illus*)
improved chemical agent monitor (ICAM), M22, 13-6 (*illus*)
improved first-aid kit (IFAK), 3-23 (*illus*)

improved M72 LAW, 10-11 (*illus*)
improvised explosive device (IED), 15-10
  actions on finding, 15-18
  components, 15-14 (*illus*)
  emplaced, with initiator, 5-6 (*illus*)
  transmitters and receivers, 5-15 (*illus*)
individual
  fighting position, 6-5 (*illus*), 8-13 (*illus*)
  movement techniques, 7-1
  techniques, 5-8
initiators
  command detonated, victim activated, with timer, 15-13 (*illus*)
  common objects as, 15-15 (*illus*)
  modernized demolition initiator, 14-10 (*illus*)
  unexploded rounds as, 15-16 (*illus*)
injured limb, elevation of, 3-18 (*illus*)

J

Javelin, 10-14 (*illus*)
  fighting position, 6-25 (*illus*)
jaw thrust technique, 3-5 (*illus*)
jungle, 4-5

K

knot toward mine, 14-16 (*illus*)

L

lanes, 14-16 (*illus*)
Law of Land Warfare, 1-4
legal assistance, 2-1
lifesaving measures (first aid), 3-2
limited visibility observation, 9-9
low and high crawl, 7-2 (*illus*)
lower-level entry techniques, 8-7 (*illus*)

M

M100 Sorbent Decontamination System, 13-15 (*illus*)

Index

M-131 modular pack mine system (MOPMS), 14-6 (*illus*)
M136 AT4, 10-12 (*illus*)
M141 BDM, 10-13 (*illus*)
M16A2/A3 rifle
 battlesight zero, 10-18 (*illus*)
 rifle mechanical zero, 10-17 (*illus*)
M16A2 rifle, 10-4 (*illus*)
M16A4 and M4 carbine rifle mechanical zero, 10-17 (*illus*)
M16A4 rifle battlesight zero, 10-18 (*illus*)
M18A1 antipersonnel mine, 14-3 (*illus*), 14-4 (*illus*)
M2 .50 caliber machine gun with M3 tripod mount, 10-9 (*illus*)
M203 grenade launcher, 10-6 (*illus*)
M21 antitank mine and components, 14-8 (*illus*)
M22
 automatic chemical agent decontaminating apparatus (ACADA), 13-5 (*illus*)
 improved chemical agent monitor (ICAM), 13-6 (*illus*)
M240B machine gun, 10-8 (*illus*)
M249 squad automatic weapon (SAW), 10-7 (*illus*)
M256 chemical agent detector kit, 13-8 (*illus*)
M291 skin decontaminating kit, 13-15 (*illus*)
M295 equipment decontamination kit, 13-15 (*illus*)
M3 tripod mount, 10-9 (*illus*)
M40A1/A2 protective mask, 13-10 (*illus*)
M42A2 CVC protective mask, 13-10 (*illus*)
M433 HEDP grenade, 8-10 (*illus*)
M4 carbine, 10-5 (*illus*)
 battlesight zero, 10-19 (*illus*)
M68 close combat optic, 10-26 (*illus*)

M72 LAW, improved, 10-11 (*illus*)
M7 bandoleer, 14-3 (*illus*)
M81 fuse igniter
 with the M14 time fuse delay, 14-12 (*illus*), 14-13 (*illus*)
 with the M9 holder, 14-13 *illus*)
M8 chemical agent detector paper, 13-6 (*illus*)
M9
 chemical agent detector paper, 13-7 (*illus*)
 pistol, 10-3 (*illus*)
machine gun
 emplacement in doorway, 8-18 (*illus*)
 fighting position, 6-23 (*illus*)
 M240B, 10-8 (*illus*)
 M2 .50 caliber, with M3 tripod mount, 10-9 (*illus*)
 MK 19 grenade, Mod 3, 10-10 (*illus*)
main charge (explosives), 15-12 (*illus*)
man-made cover, 5-2 (*illus*), 6-1 (*illus*)
Mark I nerve agent antidote kit (NAAK), 13-2 (*illus*)
maximum engagement line (MEL), 6-34 (*illus*)
MBITR. *See* multiband intrateam radio
mechanical zero, 10-17 (*illus*)
mental health and morale, 3-41
message format, 11-4
messengers, 11-1
mil-relation formula, 9-17 (*illus*)
mines, 14-1
 marked, 14-17 (*illus*)
 probe, 14-15 (*illus*)
misfires, 14-14
 procedures, 10-31
mission-oriented protective posture (MOPP), 13-11 (*illus*)
MK 19 grenade machine gun, Mod 3, 10-10 (*illus*)
MK7 antipersonnel obstacle-breaching system (APOBS), 14-20 (*illus*)

modernized demolition initiator (MDI), 14-9 (*illus*)
Modular Pack Mine System (MOPMS), 14-7 (*illus*)
movement, 5-5, 8-1. *See also* camouflage, individual, 7-1
 on vehicles, 7-9
multiband intrateam radio (MBITR), AN/PRC-148, 11-12 (*illus*)

N

natural cover, 5-1 (*illus*)
neck drag, 3-34 (*illus*)
nerve agent antidote kit (NAAK), Mark I, 13-2 (*illus*)
net, 11-8
 types of, 11-4
Nine-Line UXO Incident Report, 15-17 (*illus*)
nuclear weapons, 13-20

O

observation, 9-8
 limited visibility, 9-9
obstacles, 14-14
off-center viewing, 9-11 (*illus*)
offset symbols, 10-25 (*illus*)
one-man
 fighting position, 6-20
 lift technique, 8-8 (*illus*)
open areas, avoiding, 8-1
operational environment, 1-2
operation on a net, 11-8
operations security (OPSEC), 9-8
ordnance, dropped, 15-1
outlines and shadows, 5-5
overhead cover, 6-17 (*illus*)
 built down, 6-18 (*illus*)
 construction, 6-11 (*illus*)
 stringer placement, 6-12 (*illus*)

P

personal predeployment checklist, 2-2 (*illus*)
personal weapon, 2-3
pie-ing a corner, 8-4 (*illus*)
positions, 5-5. *See also* camouflage
 digging (side view), 6-12 (*illus*)
potential indicators, 9-2 (*illus*)

Index

precedence of reports, 11-4
predeployment, 2-1
preparation of M81 fuse igniter, 14-12 (*illus*)
prepared fighting position, 8-16
pressure
 digital, 3-20 (*illus*)
 direct manual, 3-18 (*illus*)
primary sector, 6-27 (*illus*)
priming. *See also* C4
 bangalore torpedo, 14-19 (*illus*)
principal direction of fire (PDF), 6-28 (*illus*)
projected ordnance, 15-6
prone position (hasty), 6-6 (*illus*)
protective cover, 5-3 (*illus*)
protective masks, 13-10 (*illus*)
prowords, 11-5
pulse, 3-8 (*illus*)

Q

questioning, forms of, 9-3

R

radiac sets, 13-21 (*illus*)
radiological weapons, 13-18
radios, 11-3, 11-12
radiotelephone procedures, 11-4
range
 card, 6-26, 6-29 through 6-31 (*illus*)
 estimation, 9-15
rapid/slow scan pattern, 9-9 (*illus*)
ready positions, 10-35 (*illus*)
reference points, 6-33 (*illus*), 10-24 (*illus*)
reflexive fire, 10-34
report levels, 9-4
rescue breathing, 3-7 (*illus*)
resistance, 12-6
revetments, 6-13 (*illus*)
rifle battlesight zero, 10-18 (*illus*)
room-clearing techniques, 8-11

rules, 11-4
rush, 7-3 (*illus*)

S

safety, 10-1
SALUTE format, 9-5 (*illus*)
SAW. *See* M249 squad automatic weapon
scanning pattern, 9-11 (*illus*)
search, detailed, 9-9 (*illus*)
sectors and fields of fire, 6-4. *See also* fighting positions
SERE. *See* survival, evasion, resistance, and escape
shelter locations, 12-5 (*illus*)
shot group adjustment, 10-29 (*illus*)
shot groups, 10-20
signal operating instructions (SOI), 11-9
skin decontaminating kit, M291, 13-15 (*illus*)
smoke grenades, 15-9 (*illus*)
soldier in arctic camouflage, 5-6 (*illus*)
Sorbent Decontamination System, M100, 13-15 (*illus*)
squad automatic weapon, 10-7 (*illus*)
start point, 10-24 (*illus*)
storage compartments, 6-14 (*illus*)
strike zone shot group adjustment, 10-29 (*illus*)
stringer placement, overhead cover, 6-12 (*illus*)
supporting carry, 3-33 (*illus*)
survival, evasion, resistance, and escape (SERE), 12-1

T

target reference points, 6-33 (*illus*)
telephone equipment, 11-14
thermal weapon sight zeroing adjustments, 10-28 (*illus*)

thrown ordnance (hand grenades), 15-8
time periods, 11-11 (*illus*)
tongue jaw lift, 3-13 (*illus*)
tourniquet, 3-24 (*illus*)
two-hand seat carry, 3-38 (*illus*)
two-man
 fighting position, 6-7 (*illus*) through 6-13
 fore and aft carry, 3-37 (*illus*)
 pull technique, 8-8 (*illus*)
 support carry, 3-36 (*illus*)

U

unexploded ordnance (UXO), 15-1
 actions, 15-17
 initiator, used as, 15-16 (*illus*)
 nine-line report, 15-17
urban areas, 8-1

V

vehicle IED capacities and danger zones, 15-11 (*illus*)
visual signals, 11-3

W

wall, crossing, 8-3 (*illus*)
warrior culture, 1-4
warrior drills, 1-5 (*illus*)
weapon reference point, 6-35 (*illus*)
weapons, 10-2
windows, movement past, 8-2 (*illus*)
wire, 11-1

Z

zeroing mark, 10-22 (*illus*)

TC 3-21.75 (FM 3-21.75)
13 August 2013

By Order of the Secretary of the Army:

RAYMOND T. ODIERNO
General, United States Army
Chief of Staff

Official:

GERALD B. O'KEEFE
Administrative Assistant to the
Secretary of the Army
1322701

Distribution:

Active Army, Army National Guard, and United States Army Reserve: Not to be distributed; electronic media only.

PIN: 103242-00

www.ingramcontent.com/pod-product-compliance
Lightning Source LLC
Chambersburg PA
CBHW050049230526
45470CB00004B/1456